Renewable Energy: An Introduction

Renewable Energy: An Introduction

Editor: Dylan Carter

Callisto Reference

www.callistoreference.com

Callisto Reference,
118-35 Queens Blvd., Suite 400,
Forest Hills, NY 11375, USA

Visit us on the World Wide Web at:
www.callistoreference.com

ISBN: 978-1-64116-010-0 (Hardback)

Printed in China.

Cataloging-in-Publication Data

Renewable energy : an introduction / edited by Dylan Carter.
p. cm.
Includes bibliographical references and index.
ISBN 978-1-64116-010-0
1. Renewable energy sources. 2. Power resources. I. Carter, Dylan.
TJ808. R46 2018
333.794--dc23

Table of Contents

Preface

Those sources of energy which are naturally replenished are called renewable resources. The energy derived from them is known as renewable energy. It is an environment-friendly source of energy. It refers to the energy derived from natural resources, like sun, water, wind, etc. Some of the most successful forms of renewable energy are solar power, hydropower, wind energy, etc. The varied aspects of renewable energy along with technological progress that have future implications are glanced at in this book. The topics in the text are apt for the readers as they provide thorough insights about this vast field. This textbook is meant for students who are looking for an elaborate reference guide on renewable energy.

A short introduction to every chapter is written below to provide an overview of the content of the book:

Chapter 1 - The energy derived from falling water is known as hydropower. It is mainly used to generate electricity. The chapter on hydroenergy offers an insightful focus, keeping in mind the complex subject matter; **Chapter 2 -** Wind power is the conversion of airflow into electric power. Wind power is a renewable source of energy that does not produce any greenhouse gases. The topics discussed in the chapter are of great importance to broaden the existing knowledge on wind and wave power; **Chapter 3 -** Solar energy is generated by technologies like solar thermal energy, solar heating, solar architecture and photovoltaics. Solar energy is an appealing source of electricity as it is renewable and economical. This chapter has been carefully written to provide an easy understanding of the varied facets of solar energy; **Chapter 4 -** Biomass is energy that is generated by burning organic matter. Some of the methods used to convert biomass are chemical, biochemical and thermal conversions. Anaerobic digestion, biofuel and biogas are some of the topics discussed. This chapter is an overview of the subject matter incorporating all the major aspects of bioenergy; **Chapter 5 -** Hybrid renewable systems are becoming very popular as they provide benefits of renewable energy to isolated areas. In order to completely understand hybrid system, it is necessary to understand the processes related to it. The following chapter elucidates the varied processes and mechanisms associated with this area of study.

Finally, I would like to thank my fellow scholars who gave constructive feedback and my family members who supported me at every step.

Editor

1

Basics of Hydroenergy

The energy derived from falling water is known as hydropower. It is mainly used to generate electricity. The chapter on hydroenergy offers an insightful focus, keeping in mind the complex subject matter.

Hydropower

Hydro-electric power is generated by the flow of water through turbine, turning the blades of the turbine. A generator shaft connected to this turbine also turns and hence generates electricity. Following figure shows how hydro-electric power is generated:

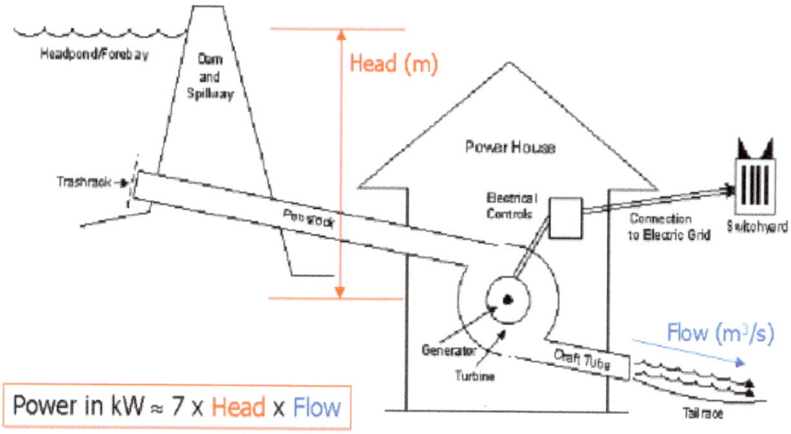

The main components of a hydel power plant are:

1. Dam/Reservoir/Large buffer tank
2. Penstock
3. Power House
   a. Turbines
   b. Generators
   c. Step-up Transformers

Depending on the capacity, hydel power plants are divided into the following categories:

Category	Capacity	Application
Large Hydel Plant	50 MW to 1000 MW	Large Cities
Small Hydel Plant	1 MW to 50 MW	Small cities to Towns
Mini Hydel Plant	100 kW to 1000 kW	Towns
Micro Hydel Plant	< 100 kW	Rural community
Pico Hydel Plant	< 5 kW	Individual home

Hydel plants have an efficiency of 75%. The power delivered is given by the following expression:

$Power\_Delivered = 7.\mathrm{H}.\frac{dQ}{dt}$ Kilo watts, where

H = Head in meters

DQ/dt = Rate of discharge in m^3/s.

In the figure we see that the turbine is coupled to a generator for generating electrical power. The generator can be of any of the following types:

- Permanent magnet DC generator (PMDC)
- Alternator (Synchronous Generator)
- Induction Generator
- Synchronous reluctance Generator

To select the best among the listed options, we need to know the requirement of a generator. Following list gives the requirement:

1. Rugged and easy to maintain
2. Simple to fabricate
3. High efficiency
4. Fail safe or "should not fail at all"
5. Sinusoidal output
6. Good voltage regulation
7. Cost effective for given power
8. Ease of servicing/operation
9. Safety
10. Reliability

When we try to match the requirements to the types of generators, Induction generator fits the bill better than others and hence, this is the type normally used for power generation. The only drawback with induction generators is its poor voltage regulation. To improve the voltage regulation, normally load governors are used in parallel to the actual load.

Hydraulic Energy

The power that is obtainable from a stream depends primarily on two factors. These are,

1. Head
2. Flow

The head is the vertical height from the top of the penstock to the bottom of the penstock. Larger the head, higher will be the speed of the turbine and larger will be the power output.

The flow is the volume of water which flows in one second.

$Energy = m.\mathrm{g}.\mathrm{h}$ Joules, where

m = mass of water = Kg

g = acceleration due to gravity = 9.81 m/s^2

h = height = meter

$Energy = \rho.\mathrm{Q}.\mathrm{g}.\mathrm{h}$ Joules, where

ρ = density of water = Kg/m^3

Q = discharge = m^3

g = acceleration due to gravity = 9.81 m/s^2

h = height = meter

$Power = \rho.\frac{dQ}{dt}.g.h$ watts, where

ρ = density of water = Kg/m^3

dQ/dt = rate of discharge = m^3/s

g = acceleration due to gravity = 9.81 m/s^2

h = height = meter

1 Kilowatt-Hour = 1000 watts x 3600 seconds = 3.6e6 watt-second = 3.6e6 Joules

1 m^3 = 1000 liters

Considering an efficiency of 70 to 75%, the hydraulic power is given by

$$power\_Delivered \approx 7 \cdot H \cdot \frac{dQ}{dt} Kilo\,watts, where$$

h = Head in meters

dQ/dt = Rate of discharge in m^3/s.

Layouts

As the physical layout of the micro hydel plant will affect the power output, cost, ruggedness etc., it is worth to consider the following options in choosing the plant layout.

Layout decision is based on what factors?

Locations of homes: The distance of the homes from the micro hydel plant will affect the cost of the overall plant.

Water rights: Water may be used downstream by many others for many purposes. The water usage has to be checked with everyone affected and negotiated right at the beginning of the planning process. Typical questions to be asked are – whose land will be used for the plant? Who else uses the water and for what purpose? What is the agreement on water sharing? etc.

Power generation: The generated power depends on the head and flow of the water. These are both affected by the physical layout.

Cost: The major cost factor in the plant is the penstock and the distribution system. The challenge is to keep the penstock and distribution system as short as possible. Both the cost and the losses increases as these become longer.

Site Survey

It is important to measure the head and the flow with reasonable accuracy to ensure that the power requirements are met. It is better to under estimate the head and the flow rather than to over estimate them to obtain conservative hydraulic power capability of the stream.

Measuring Head

Three popular methods for head measurement are:

1. Water filled plastic tube
2. Altimeter
3. Abney level

Water Filled Plastic Tube

This is the most inexpensive method of head measurement. It requires a piece of transparent plastic tube about 20m long and a diameter of 10mm. Fill the tube with water so that when the two ends are held together, the water level is about 300mm from the top. The water inside the tube will always find the same level on either side. A plastic funnel will help to pour in the water.

One person holds one end of the tube at the water level of the reservoir/forebay tank. The second person moves downhill till his eyes are in level with the water level of the fore bay tank. His end of the tube is adjusted till the water level in the tube is in level with his eyes. Now record that one reading has been taken. After this the tube is lowered such that the water level in the tube is in line with the soles of his feet.

Now the first person moves downhill till his eye level is in line with the soles of the feet of the first person. He now raises the tube till the water level is in line with his eye. Now record that a second reading has been taken. This process is repeated till the location of the turbine. The number of readings taken is summed up. This is multiplied with the average height to eye level of the two people who took the measurements. This gives the total head. This procedure should be repeated two to three times to obtain good accuracy.

A variation of this method is to connect one end of the tube with a pressure gauge. The pressure at each measuring point is recorded and the sum of the total pressures can be used to calculate the overall head.

Altimeters

Altimeters are an instrument to measure the head. Height is calculated using changes in air pressure. All that the user had to do is to record one reading at the expected fore bay location. Then he has to record the second reading at the turbine in order to determine the head. The second reading should be taken as quickly as possible to prevent any atmospheric changes to affect the reading.

Abney Level

The Abney level is a hand held sighting meter. With this method, the angle of the slope (θ) is measured. The linear distance (L) is also measured. The head is H is given by

$$H = L. \sin(\theta)$$

Two posts are driven into the ground one at the position of the proposed fore bay and the other at a position that is 20m to 30m downhill. A clear line of site between the top of the posts is required. The angle between the tops of the two posts is measured using the Abney level and the distance between them is also recorded. The head is measure between these two posts using the above relationship. Now, the first post is shifted further downhill as compared to the second post and the measurement recorded. This process is repeated till the position of the proposed turbine is reached. The heights are all added up to obtain the overall head.

Measuring Flow

The flow should be measured during the worst case condition. During the rainy season, the flow will be high and will provide high power output. However, during the dry season, the flow will be low. It should be estimated whether the load requirement will be met even during the worst case dry seasons. Therefore, the flow measurement should be done during the dry seasons. Three simple methods of flow measurements are

1. Bucket method
2. Float method
3. Salt Gulp Analysis

Bucket Method

Take a 15 liter bucket or any container with known volume. If the volume of the container is not known then it can be found out by filling the container with water from a 1litre bottle. Count the number of liter that has been added. This is giving the volume of the container.

The most challenging task in this method is to find a location in the stream where the water can be directed into the bucket. It is important as little as possible escapes. This is done by placing the bucket at the bottom of a natural narrow fall in the stream path or by building a simple weir by using a wooden channel or a corrugated sheet such that the stream water flows through this weir during measurement. Now, using a stopwatch, record the time it takes to fill the bucket. Repeat the experiment three times and take the average. The volume of the bucket divided by the average time it takes to fill the bucket gives the flow rate of the water.

Float Method

This method works well in canals and channels because this method requires the knowledge of the cross sectional area of water flowing. The speed of the water flowing is found out by using a float and timing its travel between two points in the stream.

Estimating the cross sectional area of a channel or canal is relatively easy as the cross sectional geometry is known. But in the case of a river or stream it is much more difficult to calculate the cross sectional area. To estimate the area at a particular section, measure the width of the stream at that point. Then with the help of a long pole, take the depth measurements at regular intervals across the stream where the width has been measured. Plot this depth measurement on a grid sheet and estimate the area of the cross section by calculating the number of squares of the grid sheet enclosed.

To measure the surface flow, place two markers at a distance of about 10m. Start the stopwatch when the float pass the first marker and stop it when the float passes the second marker. Repeat the experiment at least three times to obtain consistent results.

The flow is the product of the average stream area and the average velocity of the flow. Since the water moves the fastest at the surface than the other layers of the stream, the average velocity of the stream is found by incorporating a correction factor to the surface velocity. The following table gives typical correction factors that can be used.

Type of stream	Velocity correction factor (vice)
A rectangular channel with smooth sides and bed	0.85
A deep, slow moving stream	0.75
A small stream with a smooth bed	0.65
A quick, turbulent stream	0.45
A very shallow, rocky stream	0.25

The flow is now given by

$$\frac{dQ}{dt} = A.v_{surface}.vcf$$

Salt Gulp Method

This method requires more calculations than the other methods but is easier to conduct the experiment. It needs a conductivity meter and an instrument to accurately weigh salt. The flow rate is determined by measuring the speed and concentration of a cloud of salty water as it passes downstream.

A known mass of salt is mixed with some water in a bucket until it is fully dissolved. Record the mass of the salt added to the nearest gram. All the salt water solution is then tipped into the stream. The mass of salt used depends on the size of the flow. As a thumb rule, use about 25g of salt for every 5 lisper second flow.

The conductivity meter probe is placed in a fast moving area of flow 25m to 30m downstream from where the salt solution was added. The normal conductivity level of the water is recorded. This is called background conductivity. As soon as the conductivity readings begin to climb, record then every 5 seconds. Readings should be recorded until the conductivity has returned to its background level.

Now plot a graph of changing salt concentration versus time. A smooth curve with a peak value at least twice the level of the background level indicates that the experiment has been carried out reasonably well. The area under the curve must be established in order to calculate the flow rate. The equation for calculating the flow rate is

$$\frac{dQ}{dt} = \frac{Mk^{-1}}{A}$$

Where M = mass of salt

K^{-1} = conversion factor

A = area under the concentration versus time curve

The conductivity is converted into salt concentration by multiplying with a conversion factor that takes water temperature into account. It has the units of $ohm^{-1}/mg\ lt^{-1}$. The k^{-1} value is around 2 for water at 25 deg. C.

Turbine

The turbine is a device which converts the hydraulic power from the water to rotating mechanical energy. There are several turbines that have been developed for specific purposes. Turbines are classified into

1. Impulse turbines for high heads.
2. Reaction turbines for low heads.
3. Submersible propeller turbines.

Some of the turbines are Elton, cross flow, turbo, Francis, Harris etc.

Generator

Generally, induction generators and synchronous generators are used to produce the power. Induction generators are especially suited for providing electricity in remote areas because they are rugged and reliable. However, synchronous generators and dc generators have also been used for some applications.

The three phase induction motors can be used as generators. These are easily available and quite inexpensive. For most home applications, a single phase supply is required. It is required to produce a single phase supply from a three phase motor. This is done by connecting unequal amounts of excitation capacitance across the windings of the machine.

Load Control

The speed of the turbine changes when the load connected to the generator changes. Since this change of speed affects the voltage and frequency, the load on the generator must be kept constant or the flow of water through the nozzle must be adjusted. The most reliable method of controlling the load and keeping the voltage and frequency constant is by using an electronic load controller. The unused power of the induction generator is sent to a ballast or dummy load so that the total load on the generator remains constant. For example if the generator produces 1000W and the total load connected is 700W, then the remaining 300W will be dissipated in the ballast load.

Tidal Power

Tidal power or tidal energy is a form of hydropower that converts the energy obtained from tides into useful forms of power, mainly electricity.

Although not yet widely used, tidal energy has potential for future electricity generation. Tides are more predictable than the wind and the sun. Among sources of renewable energy, tidal energy has traditionally suffered from relatively high cost and limited availability of sites with sufficiently high tidal ranges or flow velocities, thus constricting its total availability. However, many recent technological developments and improvements, both in design (e.g. dynamic tidal power, tidal lagoons) and turbine technology (e.g. new axial turbines, cross flow turbines), indicate that the total availability of tidal power may be much higher than previously assumed, and that economic and environmental costs may be brought down to competitive levels.

Historically, tide mills have been used both in Europe and on the Atlantic coast of North America. The incoming water was contained in large storage ponds, and as the tide went out, it turned waterwheels that used the mechanical power it produced to mill grain. The earliest occurrences date from the Middle Ages, or even from Roman times. The process of using falling water and spinning turbines to create electricity was introduced in the U.S. and Europe in the 19th century.

The world's first large-scale tidal power plant was the Rance Tidal Power Station in France, which became operational in 1966. It was the largest tidal power station in terms of output until Sihwa Lake Tidal Power Station opened in South Korea in August 2011. The Sihwa station uses sea wall defense barriers complete with 10 turbines generating 254 MW.

Generation of Tidal Energy

Tidal power is taken from the Earth's oceanic tides. Tidal forces are periodic variations in gravitational attraction exerted by celestial bodies. These forces create corresponding motions or currents in the world's oceans. Due to the strong attraction to the oceans, a bulge in the water level is created, causing a temporary increase in sea level. When the sea level is raised, water from the middle of the ocean is forced to move toward the shorelines, creating a tide. This occurrence takes place in an unfailing manner, due to the consistent pattern of the moon's orbit around the earth. The magnitude and character of this motion reflects the changing positions of the Moon and Sun relative to the Earth, the effects of Earth's rotation, and local geography of the sea floor and coastlines.

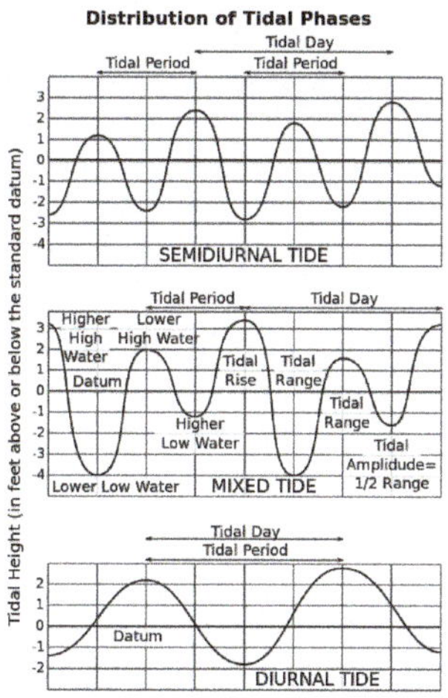

Variation of tides over a day

Tidal power is the only technology that draws on energy inherent in the orbital characteristics of the Earth–Moon system, and to a lesser extent in the Earth–Sun system. Other natural energies exploited by human technology originate directly or indirectly with the Sun, including fossil fuel, conventional hydroelectric, wind, biofuel, wave and solar energy. Nuclear energy makes use of Earth's mineral deposits of fissionable elements, while geothermal power taps the Earth's internal heat, which comes from a combination of residual heat from planetary accretion (about 20%) and heat produced through radioactive decay (80%).

A tidal generator converts the energy of tidal flows into electricity. Greater tidal variation and higher tidal current velocities can dramatically increase the potential of a site for tidal electricity generation.

Because the Earth's tides are ultimately due to gravitational interaction with the Moon and Sun and the Earth's rotation, tidal power is practically inexhaustible and classified as a renewable energy resource. Movement of tides causes a loss of mechanical energy in the Earth–Moon system: this is a result of pumping of water through natural restrictions around coastlines and consequent viscous dissipation at the seabed and in turbulence. This loss of energy has caused the rotation of the Earth to slow in the 4.5 billion years since its formation. During the last 620 million years the period of rotation of the earth (length of a day) has increased from 21.9 hours to 24 hours; in this period the Earth has lost 17% of its rotational energy. While tidal power will take additional energy from the system, the effect is negligible and would only be noticed over millions of years.

Generating Methods

The world's first commercial-scale and grid-connected tidal stream generator – SeaGen – in Strangford Lough. The strong wake shows the power in the tidal current.

Tidal power can be classified into four generating methods:

Tidal Stream Generator

Tidal stream generators (or TSGs) make use of the kinetic energy of moving water to power turbines, in a similar way to wind turbines that use wind to power turbines. Some tidal generators can be built into the structures of existing bridges or are entirely submersed, thus avoiding concerns over impact on the natural landscape. Land constrictions such as straits or inlets can create high velocities at specific sites, which can be captured with the use of turbines. These turbines can be horizontal, vertical, open, or ducted.

Tidal Barrage

Tidal barrages make use of the potential energy in the difference in height (or hydraulic head) between high and low tides. When using tidal barrages to generate power, the potential energy from a tide is seized through strategic placement of specialized dams. When the sea level rises and the tide begins to come in, the temporary increase in tidal power is channeled into a large basin behind the dam, holding a large amount of potential energy. With the receding tide, this energy is then converted into mechanical energy as the water is released through large turbines that create electrical power through the use of generators. Barrages are essentially dams across the full width of a tidal estuary.

Dynamic Tidal Power

Dynamic tidal power (or DTP) is an untried but promising technology that would exploit an interaction between potential and kinetic energies in tidal flows. It proposes that very long dams (for example: 30–50 km length) be built from coasts straight out into the sea or ocean, without enclosing an area. Tidal phase differences are introduced across the dam, leading to a significant water-level differential in shallow coastal seas – featuring strong coast-parallel oscillating tidal currents such as found in the UK, China, and Korea.

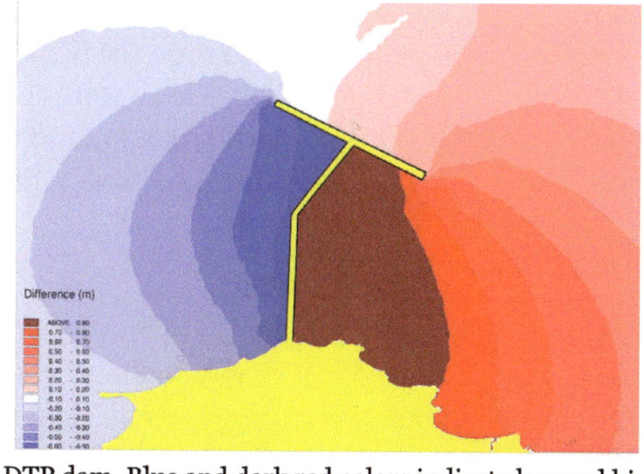

Top-down view of a DTP dam. Blue and dark red colors indicate low and high tides, respectively.

Tidal Lagoon

A new tidal energy design option is to construct circular retaining walls embedded with turbines that can capture the potential energy of tides. The created reservoirs are similar to those of tidal barrages, except that the location is artificial and does not contain a preexisting ecosystem. The lagoons can also be in double (or triple) format without pumping or with pumping that will flatten out the power output. The pumping power could be provided by excess to grid demand renewable energy from for example wind turbines or solar photovoltaic arrays. Excess renewable energy rather than being curtailed could be used and stored for a later period of time. Geographically dispersed tidal lagoons with a time delay between peak production would also flatten out peak production providing near base load production though at a higher cost than some other alternatives such as district heating renewable energy storage. The proposed Tidal Lagoon Swansea Bay in Wales, United Kingdom would be the first tidal power station of this type once built.

US and Canadian Studies in the Twentieth Century

The first study of large scale tidal power plants was by the US Federal Power Commission in 1924 which if built would have been located in the northern border area of the US state of Maine and the south eastern border area of the Canadian province of New Brunswick, with various dams, powerhouses, and ship locks enclosing the Bay of Fundy and Passamaquoddy Bay. Nothing came of the study and it is unknown whether Canada had been approached about the study by the US Federal Power Commission.

In 1956, utility Nova Scotia Light and Power of Halifax commissioned a pair of studies into the feasibility of commercial tidal power development on the Nova Scotia side of the Bay of Fundy. The two studies, by Stone & Webster of Boston and by Montreal Engineering Company of Montreal independently concluded that millions of horsepower could be harnessed from Fundy but that development costs would be commercially prohibitive at that time.

There was also a report on the international commission in April 1961 entitled "Investigation of the International Passamaquoddy Tidal Power Project" produced by both the US and Canadian Federal Governments. According to benefit to costs ratios, the project was beneficial to the US but not to Canada. A highway system along the top of the dams was envisioned as well.

A study was commissioned by the Canadian, Nova Scotian and New Brunswick governments (Reassessment of Fundy Tidal Power) to determine the potential for tidal barrages at Chignecto Bay and Minas Basin – at the end of the Fundy Bay estuary. There were three sites determined to be financially feasible: Shepody Bay (1550 MW), Cumberline Basin (1085 MW), and Cobequid Bay (3800 MW). These were never built despite their apparent feasibility in 1977.

Tidal Power Development in the UK

The world's first marine energy test facility was established in 2003 to start the development of the wave and tidal energy industry in the UK. Based in Orkney, Scotland, the European Marine Energy Centre (EMEC) has supported the deployment of more wave and tidal energy devices than at any other single site in the world. EMEC provides a variety of test sites in real sea conditions. Its grid connected tidal test site is located at the Fall of Warness, off the island of Eday, in a narrow channel which concentrates the tide as it flows between the Atlantic Ocean and North Sea. This area has a very strong tidal current, which can travel up to 4 m/s (8 knots) in spring tides. Tidal energy developers that have tested at the site include: Alstom (formerly Tidal Generation Ltd); ANDRITZ HYDRO Hammerfest; Atlantis Resources Corporation; Nautricity; OpenHydro; Scotrenewables Tidal Power; Voith. The resource could be 4 TJ per year. Elsewhere in the UK, annual energy of 50 TWh can be extracted if 25 GW capacity is installed with pivotable blades.

Current and Future Tidal Power Schemes

- The first tidal power station was the Rance tidal power plant built over a period of 6 years from 1960 to 1966 at La Rance, France. It has 240 MW installed capacity.
- 254 MW Sihwa Lake Tidal Power Plant in South Korea is the largest tidal power installation in the world. Construction was completed in 2011.
- The first tidal power site in North America is the Annapolis Royal Generating Station, Annapolis Royal, Nova Scotia, which opened in 1984 on an inlet of the Bay of Fundy. It has 20 MW installed capacity.
- The Jiangxia Tidal Power Station, south of Hangzhou in China has been operational since 1985, with current installed capacity of 3.2 MW. More tidal power is planned near the mouth of the Yalu River.
- The first in-stream tidal current generator in North America (Race Rocks Tidal Power Demonstration Project) was installed at Race Rocks on southern Vancouver Island in September 2006. The next phase in the development of this tidal current generator will be in Nova Scotia (Bay of Fundy).
- A small project was built by the Soviet Union at Kislaya Guba on the Barents Sea. It has 0.4 MW installed capacity. In 2006 it was upgraded with a 1.2MW experimental advanced orthogonal turbine.
- Jindo Uldolmok Tidal Power Plant in South Korea is a tidal stream generation scheme planned to be expanded progressively to 90 MW of capacity by 2013. The first 1 MW was installed in May 2009.

- A 1.2 MW SeaGen system became operational in late 2008 on Strangford Lough in Northern Ireland.
- The contract for an 812 MW tidal barrage near Ganghwa Island (South Korea) north-west of Incheon has been signed by Daewoo. Completion is planned for 2015.
- A 1,320 MW barrage built around islands west of Incheon is proposed by the South Korean government, with projected construction starting in 2017.
- The Scottish Government has approved plans for a 10MW array of tidal stream generators near Islay, Scotland, costing 40 million pounds, and consisting of 10 turbines – enough to power over 5,000 homes. The first turbine is expected to be in operation by 2013.
- The Indian state of Gujarat is planning to host South Asia's first commercial-scale tidal power station. The company Atlantis Resources planned to install a 50MW tidal farm in the Gulf of Kutch on India's west coast, with construction starting early in 2012.
- Ocean Renewable Power Corporation was the first company to deliver tidal power to the US grid in September, 2012 when its pilot TidGen system was successfully deployed in Cobscook Bay, near Eastport.
- In New York City, 30 tidal turbines will be installed by Verdant Power in the East River by 2015 with a capacity of 1.05MW.
- Construction of a 320 MW tidal lagoon power plant outside the city of Swansea in the UK was granted planning permission in June 2015 and work is expected to start in 2016. Once completed, it will generate over 500GWh of electricity per year, enough to power roughly 155,000 homes.
- A turbine project is being installed in Ramsey Sound in 2014.
- The largest tidal energy project entitled MeyGen (398MW) is currently in construction in the Pentland Firth in northern Scotland.
- A combination of 5 tidal stream turbines from Tocardo are placed in the Oosterscheldekering, the Netherlands, and have been operational since 2015 with a capacity of 1,2 MW.

Tidal Power Issues

Environmental Concerns

Tidal power can have effects on marine life. The turbines can accidentally kill swimming sea life with the rotating blades, although projects such as the one in Strangford feature a safety mechanism that turns off the turbine when marine animals approach. Some fish may no longer utilize the area if threatened with a constant rotating or noise-making object. Marine life is a huge factor when placing tidal power energy generators in the water and precautions are made to ensure that as many marine animals as possible will not be affected by it. The Tethys database provides access to scientific literature and general information on the potential environmental effects of tidal energy.

Tidal Turbines

The main environmental concern with tidal energy is associated with blade strike and entanglement of marine organisms as high speed water increases the risk of organisms being pushed near or through these devices. As with all offshore renewable energies, there is also a concern about how the creation of EMF and acoustic outputs may affect marine organisms. It should be noted that because these devices are in the water, the acoustic output can be greater than those created with offshore wind energy. Depending on the frequency and amplitude of sound generated by the tidal energy devices, this acoustic output can have varying effects on marine mammals (particularly those who echolocate to communicate and navigate in the marine environment, such as dolphins and whales). Tidal energy removal can also cause environmental concerns such as degrading farfield water quality and disrupting sediment processes. Depending on the size of the project, these effects can range from small traces of sediment building up near the tidal device to severely affecting nearshore ecosystems and processes.

Tidal Barrage

Installing a barrage may change the shoreline within the bay or estuary, affecting a large ecosystem that depends on tidal flats. Inhibiting the flow of water in and out of the bay, there may also be less flushing of the bay or estuary, causing additional turbidity (suspended solids) and less saltwater, which may result in the death of fish that act as a vital food source to birds and mammals. Migrating fish may also be unable to access breeding streams, and may attempt to pass through the turbines. The same acoustic concerns apply to tidal barrages. Decreasing shipping accessibility can become a socio-economic issue, though locks can be added to allow slow passage. However, the barrage may improve the local economy by increasing land access as a bridge. Calmer waters may also allow better recreation in the bay or estuary. In August 2004, a humpback whale swam through the open sluice gate of the Annapolis Royal Generating Station at slack tide, ending up trapped for several days before eventually finding its way out to the Annapolis Basin.

Tidal Lagoon

Environmentally, the main concerns are blade strike on fish attempting to enter the lagoon, acoustic output from turbines, and changes in sedimentation processes. However, all these effects are localized and do not affect the entire estuary or bay.

Corrosion

Salt water causes corrosion in metal parts. It can be difficult to maintain tidal stream generators due to their size and depth in the water. The use of corrosion-resistant materials such as stainless steels, high-nickel alloys, copper-nickel alloys, nickel-copper alloys and titanium can greatly reduce, or eliminate, corrosion damage.

Mechanical fluids, such as lubricants, can leak out, which may be harmful to the marine life nearby. Proper maintenance can minimize the amount of harmful chemicals that may enter the environment.

Fouling

The biological events that happen when placing any structure in an area of high tidal currents and high biological productivity in the ocean will ensure that the structure becomes an ideal substrate for the growth of marine organisms. In the references of the Tidal Current Project at Race Rocks in British Columbia this is documented. Several structural materials and coatings were tested by the Lester Pearson College divers to assist Clean Current in reducing fouling on the turbine and other underwater infrastructure.

Structural Health Monitoring

The high load factors resulting from the fact that water is 800 times denser than air and the predictable and reliable nature of tides compared with the wind makes tidal energy particularly attractive for Electric power generation. Condition monitoring is the key for exploiting it cost-efficiently.

Hydroelectricity

The Three Gorges Dam in Central China is the world's largest power producing facility of any kind.

Hydroelectricity is electricity produced from hydropower. In 2015 hydropower generated 16.6% of the world's total electricity and 70% of all renewable electricity, and was expected to increase about 3.1% each year for the next 25 years.

Hydropower is produced in 150 countries, with the Asia-Pacific region generating 33 percent of global hydropower in 2013. China is the largest hydroelectricity producer, with 920 TWh of production in 2013, representing 16.9 percent of domestic electricity use.

The cost of hydroelectricity is relatively low, making it a competitive source of renewable electricity. The hydro station consumes no water, unlike coal or gas plants. The average cost of electricity from a hydro station larger than 10 megawatts is 3 to 5 U.S. cents per kilowatt-hour. With a dam and reservoir it is also a flexible source of electricity since the amount produced by the station can be changed up or down very quickly to adapt to changing energy demands. Once a hydroelectric complex is constructed, the project produces no direct waste, and has a considerably lower output level of greenhouse gases than fossil fuel powered energy plants.

History

Hydropower has been used since ancient times to grind flour and perform other tasks. In the mid-1770s, French engineer Bernard Forest de Bélidor published *Architecture Hydraulique* which described vertical- and horizontal-axis hydraulic machines. By the late 19th century, the electrical generator was developed and could now be coupled with hydraulics. The growing demand for the Industrial Revolution would drive development as well. In 1878 the world's first hydroelectric power scheme was developed at Cragside in Northumberland, England by William George Armstrong. It was used to power a single arc lamp in his art gallery. The old Schoelkopf Power Station No. 1 near Niagara Falls in the U.S. side began to produce electricity in 1881. The first Edison hydroelectric power station, the Vulcan Street Plant, began operating September 30, 1882, in Appleton, Wisconsin, with an output of about 12.5 kilowatts. By 1886 there were 45 hydroelectric power stations in the U.S. and Canada. By 1889 there were 200 in the U.S. alone.

Museum Hydroelectric power plant "Under the Town" in Serbia, built in 1900.

At the beginning of the 20th century, many small hydroelectric power stations were being constructed by commercial companies in mountains near metropolitan areas. Grenoble, France held the International Exhibition of Hydropower and Tourism with over one million visitors. By 1920 as 40% of the power produced in the United States was hydroelectric, the Federal Power Act was enacted into law. The Act created the Federal Power Commission to regulate hydroelectric power stations on federal land and water. As the power stations became larger, their associated dams developed additional purposes to include flood control, irrigation and navigation. Federal funding became necessary for large-scale development and federally owned corporations, such as the Tennessee Valley Authority (1933) and the Bonneville Power Administration (1937) were created. Additionally, the Bureau of Reclamation which had begun a series of western U.S. irrigation projects in the early 20th century was now constructing large hydroelectric projects such as the 1928 Hoover Dam. The U.S. Army Corps of Engineers was also involved in hydroelectric development, completing the Bonneville Dam in 1937 and being recognized by the Flood Control Act of 1936 as the premier federal flood control agency.

Hydroelectric power stations continued to become larger throughout the 20th century. Hydropower was referred to as *white coal* for its power and plenty. Hoover Dam's initial 1,345 MW power station was the world's largest hydroelectric power station in 1936; it was eclipsed by the 6809 MW Grand Coulee Dam in 1942. The Itaipu Dam opened in 1984 in South America as the largest, producing 14,000 MW but was surpassed in 2008 by the Three Gorges Dam in China at 22,500 MW. Hydroelectricity would eventually supply some countries, including Norway, Democratic Republic of the Congo, Paraguay and Brazil, with over 85% of their electricity. The United

States currently has over 2,000 hydroelectric power stations that supply 6.4% of its total electrical production output, which is 49% of its renewable electricity.

Future Potential

The technical potential for the growth of hydropower around the world are, 71% Europe, 75% North America, 79% South America, 95% Africa, 95% Middle East, 82% Asia Pacific. The political realities of new reservoirs in western countries, economic limitations in the third world and the lack of a transmission system in undeveloped areas, result in the possibility of developing 25% of the remaining potential before 2050, with the bulk of that being in the Asia Pacific area. A few countries are highly developed and have very little room for growth, Switzerland 12% and Mexico 20%.

Generating Methods

Turbine row at El Nihuil II Power Station in Mendoza, Argentina

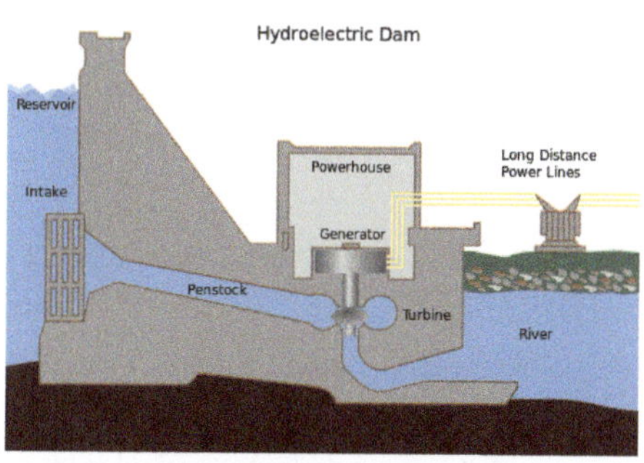

Cross section of a conventional hydroelectric dam.

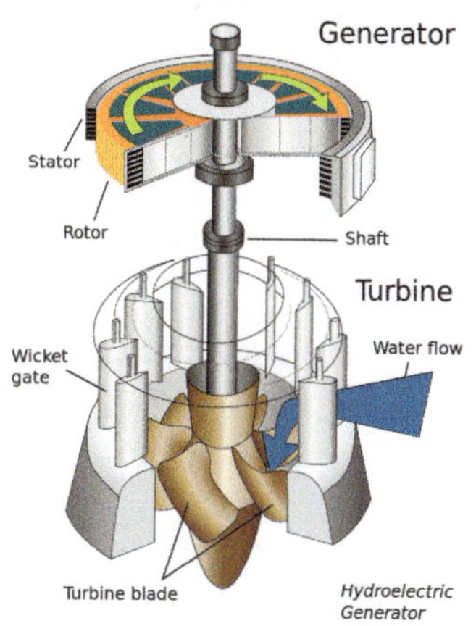

A typical turbine and generator

Conventional (Dams)

Most hydroelectric power comes from the potential energy of dammed water driving a water turbine and generator. The power extracted from the water depends on the volume and on the difference in height between the source and the water's outflow. This height difference is called the head. A large pipe (the "penstock") delivers water from the reservoir to the turbine.

Pumped-storage

This method produces electricity to supply high peak demands by moving water between reservoirs at different elevations. At times of low electrical demand, the excess generation capacity is used to pump water into the higher reservoir. When the demand becomes greater, water is released back into the lower reservoir through a turbine. Pumped-storage schemes currently provide the most commercially important means of large-scale grid energy storage and improve the daily capacity factor of the generation system. Pumped storage is not an energy source, and appears as a negative number in listings.

Run-of-the-river

Run-of-the-river hydroelectric stations are those with small or no reservoir capacity, so that only the water coming from upstream is available for generation at that moment, and any oversupply must pass unused. A constant supply of water from a lake or existing reservoir upstream is a significant advantage in choosing sites for run-of-the-river. In the United States, run of the river hydropower could potentially provide 60,000 megawatts (80,000,000 hp) (about 13.7% of total use in 2011 if continuously available).

Tide

A tidal power station makes use of the daily rise and fall of ocean water due to tides; such sources are highly predictable, and if conditions permit construction of reservoirs, can also be dispatchable to generate power during high demand periods. Less common types of hydro schemes use water's kinetic energy or undammed sources such as undershot water wheels. Tidal power is viable in a relatively small number of locations around the world. In Great Britain, there are eight sites that could be developed, which have the potential to generate 20% of the electricity used in 2012.

Sizes, Types and Capacities of Hydroelectric Facilities

Large facilities

Large-scale hydroelectric power stations are more commonly seen as the largest power producing facilities in the world, with some hydroelectric facilities capable of generating more than double the installed capacities of the current largest nuclear power stations.

Although no official definition exists for the capacity range of large hydroelectric power stations, facilities from over a few hundred megawatts are generally considered large hydroelectric facilities.

Currently, only four facilities over 10 GW (10,000 MW) are in operation worldwide.

Rank	Station	Country	Location	Capacity (MW)
1.	Three Gorges Dam	China	30°49′15″N 111°00′08″E30.82083°N 111.00222°E	22,500
2.	Itaipu Dam	Brazil Paraguay	25°24′31″S 54°35′21″W25.40861°S 54.58917°W	14,000
3.	Xiluodu Dam	China	28°15′35″N 103°38′58″E28.25972°N 103.64944°E	13,860
4.	Guri Dam	Venezuela	07°45′59″N 62°59′57″W7.76639°N 62.99917°W	10,200

Small

Small hydro is the development of hydroelectric power on a scale serving a small community or industrial plant. The definition of a small hydro project varies but a generating capacity of up to 10 megawatts (MW) is generally accepted as the upper limit of what can be termed small hydro. This may be stretched to 25 MW and 30 MW in Canada and the United States. Small-scale hydroelectricity production grew by 28% during 2008 from 2005, raising the total world small-hydro capacity to 85 GW. Over 70% of this was in China (65 GW), followed by Japan (3.5 GW), the United States (3 GW), and India (2 GW).

A micro-hydro facility in Vietnam

Pico hydroelectricity in Mondulkiri, Cambodia

Small hydro stations may be connected to conventional electrical distribution networks as a source of low-cost renewable energy. Alternatively, small hydro projects may be built in isolated areas that would be uneconomic to serve from a network, or in areas where there is no national electrical distribution network. Since small hydro projects usually have minimal reservoirs and civil construction work, they are seen as having a relatively low environmental impact compared to large hydro. This decreased environmental impact depends strongly on the balance between stream flow and power production.

Micro

Micro hydro is a term used for hydroelectric power installations that typically produce up to 100 kW of power. These installations can provide power to an isolated home or small community, or are sometimes connected to electric power networks. There are many of these installations around the world, particularly in developing nations as they can provide an economical source of energy without purchase of fuel. Micro hydro systems complement photovoltaic solar energy systems because in many areas, water flow, and thus available hydro power, is highest in the winter when solar energy is at a minimum.

Pico

Pico hydro is a term used for hydroelectric power generation of under 5 kW. It is useful in small, remote communities that require only a small amount of electricity. For example, to power one or two fluorescent light bulbs and a TV or radio for a few homes. Even smaller turbines of 200-300W may power a single home in a developing country with a drop of only 1 m (3 ft). A Pico-hydro setup is typically run-of-the-river, meaning that dams are not used, but rather pipes divert some of the flow, drop this down a gradient, and through the turbine before returning it to the stream.

Underground

An underground power station is generally used at large facilities and makes use of a large natural height difference between two waterways, such as a waterfall or mountain lake. An underground tunnel is constructed to take water from the high reservoir to the generating hall built in an underground cavern near the lowest point of the water tunnel and a horizontal tailrace taking water away to the lower outlet waterway.

Measurement of the tailrace and forebay rates at the Limestone Generating Station in Manitoba, Canada.

Calculating Available Power

A simple formula for approximating electric power production at a hydroelectric station is: $P = \rho hrgk$, where

- P is Power in watts,
- ρ is the density of water (~1000 kg/m³),
- h is height in meters,
- r is flow rate in cubic meters per second,
- g is acceleration due to gravity of 9.8 m/s²,
- k is a coefficient of efficiency ranging from 0 to 1. Efficiency is often higher (that is, closer to 1) with larger and more modern turbines.

Annual electric energy production depends on the available water supply. In some installations, the water flow rate can vary by a factor of 10:1 over the course of a year.

Properties

Advantages

The Ffestiniog Power Station can generate 360 MW of electricity within 60 seconds of the demand arising.

Flexibility

Hydropower is a flexible source of electricity since stations can be ramped up and down very quickly to adapt to changing energy demands. Hydro turbines have a start-up time of the order of a few minutes. It takes around 60 to 90 seconds to bring a unit from cold start-up to full load; this is much shorter than for gas turbines or steam plants. Power generation can also be decreased quickly when there is a surplus power generation. Hence the limited capacity of hydropower units is not generally used to produce base power except for vacating the flood pool or meeting downstream needs. Instead, it serves as backup for non-hydro generators.

Low Cost/High Value Power

The major advantage of conventional hydroelectric dams with reservoirs is their ability to store water at low cost for dispatch later as high value clean electricity. The average cost of electricity from a hydro station larger than 10 megawatts is 3 to 5 U.S. cents per kilowatt-hour. When used as peak power to meet demand, hydroelectricity has a higher value than base power and a much higher value compared to intermittent energy sources.

Hydroelectric stations have long economic lives, with some plants still in service after 50–100 years. Operating labor cost is also usually low, as plants are automated and have few personnel on site during normal operation.

Where a dam serves multiple purposes, a hydroelectric station may be added with relatively low construction cost, providing a useful revenue stream to offset the costs of dam operation. It has been calculated that the sale of electricity from the Three Gorges Dam will cover the construction costs after 5 to 8 years of full generation. Additionally, some data shows that in most countries large hydropower dams will be too costly and take too long to build to deliver a positive risk adjusted return, unless appropriate risk management measures are put in place.

Suitability for Industrial Applications

While many hydroelectric projects supply public electricity networks, some are created to serve specific industrial enterprises. Dedicated hydroelectric projects are often built to provide the substantial amounts of electricity needed for aluminium electrolytic plants, for example. The Grand Coulee Dam switched to support Alcoa aluminium in Bellingham, Washington, United States for American World War II airplanes before it was allowed to provide irrigation and power to citizens (in addition to aluminium power) after the war. In Suriname, the Brokopondo Reservoir was constructed to provide electricity for the Alcoa aluminium industry. New Zealand's Manapouri Power Station was constructed to supply electricity to the aluminium smelter at Tiwai Point.

Reduced CO_2 Emissions

Since hydroelectric dams do not use fuel, power generation does not produce carbon dioxide. While carbon dioxide is initially produced during construction of the project, and some methane is given off annually by reservoirs, hydro normally has the lowest lifecycle greenhouse gas emissions for power generation. Compared to fossil fuels generating an equivalent amount of electricity, hydro displaced three billion tonnes of CO2 emissions in 2011. One measurement of greenhouse gas related and other externality comparison between energy sources can be found in the ExternE project by the Paul Scherrer Institute and the University of Stuttgart which was funded by the European Commission. According to that study, hydroelectricity produces the least amount of greenhouse gases and externality of any energy source. Coming in second place was wind, third was nuclear energy, and fourth was solar photovoltaic. The low greenhouse gas impact of hydroelectricity is found especially in temperate climates. The above study was for local energy in Europe; presumably similar conditions prevail in North America and Northern Asia, which all see a regular, natural freeze/thaw cycle (with associated seasonal plant decay and regrowth). Greater greenhouse gas emission impacts are found in the tropical regions because the reservoirs of power stations in tropical regions produce a larger amount of methane than those in temperate areas.

Other Uses of the Reservoir

Reservoirs created by hydroelectric schemes often provide facilities for water sports, and become tourist attractions themselves. In some countries, aquaculture in reservoirs is common. Multi-use dams installed for irrigation support agriculture with a relatively constant water supply. Large hydro dams can control floods, which would otherwise affect people living downstream of the project.

Disadvantages

Ecosystem Damage and Loss of Land

Hydroelectric power stations that use dams would submerge large areas of land due to the requirement of a reservoir. Merowe Dam in Sudan.

Large reservoirs associated with traditional hydroelectric power stations result in submersion of extensive areas upstream of the dams, sometimes destroying biologically rich and productive lowland and riverine valley forests, marshland and grasslands. Damming interrupts the flow of rivers and can harm local ecosystems, and building large dams and reservoirs often involves displacing people and wildlife. The loss of land is often exacerbated by habitat fragmentation of surrounding areas caused by the reservoir.

Hydroelectric projects can be disruptive to surrounding aquatic ecosystems both upstream and downstream of the plant site. Generation of hydroelectric power changes the downstream river environment. Water exiting a turbine usually contains very little suspended sediment, which can lead to scouring of river beds and loss of riverbanks. Since turbine gates are often opened intermittently, rapid or even daily fluctuations in river flow are observed.

Siltation and Flow Shortage

When water flows it has the ability to transport particles heavier than itself downstream. This has a negative effect on dams and subsequently their power stations, particularly those on rivers or within catchment areas with high siltation. Siltation can fill a reservoir and reduce its capacity to control floods along with causing additional horizontal pressure on the upstream portion of the dam. Eventually, some reservoirs can become full of sediment and useless or over-top during a flood and fail.

Changes in the amount of river flow will correlate with the amount of energy produced by a dam.

Lower river flows will reduce the amount of live storage in a reservoir therefore reducing the amount of water that can be used for hydroelectricity. The result of diminished river flow can be power shortages in areas that depend heavily on hydroelectric power. The risk of flow shortage may increase as a result of climate change. One study from the Colorado River in the United States suggest that modest climate changes, such as an increase in temperature in 2 degree Celsius resulting in a 10% decline in precipitation, might reduce river run-off by up to 40%. Brazil in particular is vulnerable due to its heavy reliance on hydroelectricity, as increasing temperatures, lower water flow and alterations in the rainfall regime, could reduce total energy production by 7% annually by the end of the century.

Methane Emissions (from Reservoirs)

The Hoover Dam in the United States is a large conventional dammed-hydro facility, with an installed capacity of 2,080 MW.

Lower positive impacts are found in the tropical regions, as it has been noted that the reservoirs of power plants in tropical regions produce substantial amounts of methane. This is due to plant material in flooded areas decaying in an anaerobic environment, and forming methane, a greenhouse gas. According to the World Commission on Dams report, where the reservoir is large compared to the generating capacity (less than 100 watts per square metre of surface area) and no clearing of the forests in the area was undertaken prior to impoundment of the reservoir, greenhouse gas emissions from the reservoir may be higher than those of a conventional oil-fired thermal generation plant.

In boreal reservoirs of Canada and Northern Europe, however, greenhouse gas emissions are typically only 2% to 8% of any kind of conventional fossil-fuel thermal generation. A new class of underwater logging operation that targets drowned forests can mitigate the effect of forest decay.

Relocation

Another disadvantage of hydroelectric dams is the need to relocate the people living where the reservoirs are planned. In 2000, the World Commission on Dams estimated that dams had physically displaced 40-80 million people worldwide.

Failure Risks

Because large conventional dammed-hydro facilities hold back large volumes of water, a failure due to poor construction, natural disasters or sabotage can be catastrophic to downriver settlements and infrastructure.

During Typhoon Nina in 1975 Banqiao Dam failed in Southern China when more than a year's worth of rain fell within 24 hours. The resulting flood resulted in the deaths of 26,000 people, and another 145,000 from epidemics. Millions were left homeless.

The creation of a dam in a geologically inappropriate location may cause disasters such as 1963 disaster at Vajont Dam in Italy, where almost 2,000 people died.

The Malpasset Dam failure in Fréjus on the French Riviera (Côte d'Azur), southern France, collapsed on December 2, 1959, killing 423 people in the resulting flood.

Smaller dams and micro hydro facilities create less risk, but can form continuing hazards even after being decommissioned. For example, the small earthen embankment Kelly Barnes Dam failed in 1977, twenty years after its power station was decommissioned; causing 39 deaths.

Comparison and Interactions with other Methods of Power Generation

Hydroelectricity eliminates the flue gas emissions from fossil fuel combustion, including pollutants such as sulfur dioxide, nitric oxide, carbon monoxide, dust, and mercury in the coal. Hydroelectricity also avoids the hazards of coal mining and the indirect health effects of coal emissions.

Nuclear Power

Compared to nuclear power, hydroelectricity construction requires altering large areas of the environment while a nuclear power station has a small footprint, and hydro-powerstation failures have caused tens of thousands of more deaths than any nuclear station failure. The creation of Garrison Dam, for example, required Native American land to create Lake Sakakawea, which has a shoreline of 1,320 miles, and caused the inhabitants to sell 94% of their arable land for $7.5 million in 1949.

However, nuclear power is relatively inflexible; although nuclear power can reduce its output reasonably quickly. Since the cost of nuclear power is dominated by its high infrastructure costs, the cost per unit energy goes up significantly with low production. Because of this, nuclear power is mostly used for baseload. By way of contrast, hydroelectricity can supply peak power at much lower cost. Hydroelectricity is thus often used to complement nuclear or other sources for load following.

Wind Power

Wind power goes through predictable variation by season, but is intermittent on a daily basis. Maximum wind generation has little relationship to peak daily electricity consumption, the wind may peak at night when power isn't needed or be still during the day when electrical demand is highest. Occasionally weather patterns can result in low wind for days or weeks at a time, a hydroelectric reservoir capable of storing weeks of output is useful to balance generation on the grid. Peak wind power can be offset by minimum hydropower and minimum wind can be offset with maximum hydropower. In this way the easily regulated character of hydroelectricity is used to compensate for the intermittent nature of wind power. Conversely, in some cases wind power can be used to spare water for later use in dry seasons.

In areas that do not have hydropower, pumped storage serves a similar role, but at a much higher cost and 20% lower efficiency. An example of this is Norway's trading with Sweden, Denmark, the Netherlands and possibly Germany or the UK in the future. Norway is 98% hydropower, while it's flatland neighbors are installing wind power.

World Hydroelectric Capacity

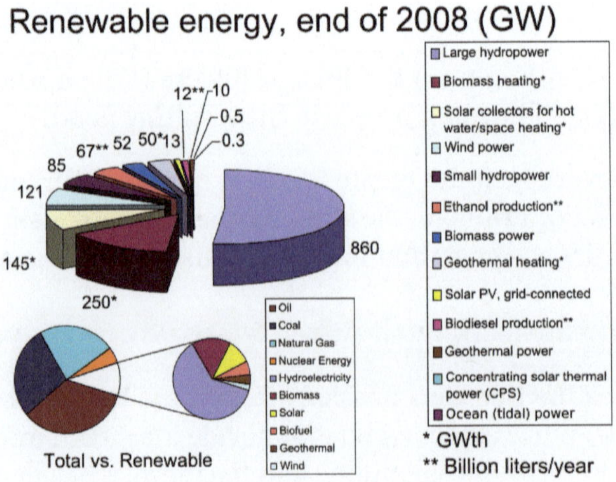

World renewable energy share (2008)

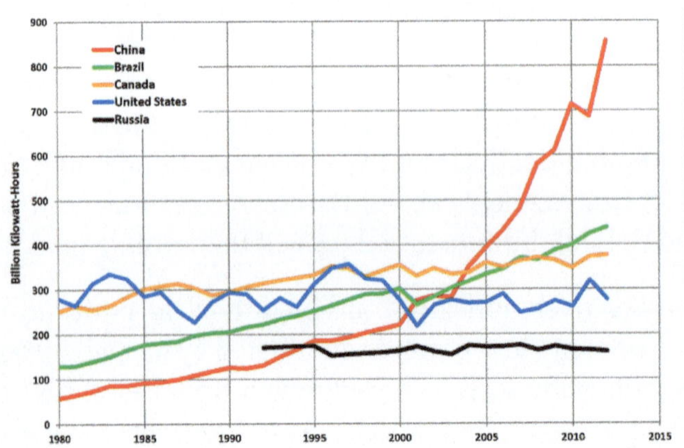

Trends in the top five hydroelectricity-producing countries

The ranking of hydro-electric capacity is either by actual annual energy production or by installed capacity power rating. In 2015 hydropower generated 16.6% of the worlds total electricity and 70% of all renewable electricity. Hydropower is produced in 150 countries, with the Asia-Pacific region generated 32 percent of global hydropower in 2010. China is the largest hydroelectricity producer, with 721 terawatt-hours of production in 2010, representing around 17 percent of domestic electricity use. Brazil, Canada, New Zealand, Norway, Paraguay, Austria, Switzerland, and Venezuela have a majority of the internal electric energy production from hydroelectric power. Paraguay produces 100% of its electricity from hydroelectric dams, and exports 90%

of its production to Brazil and to Argentina. Norway produces 98–99% of its electricity from hydroelectric sources.

A hydro-electric station rarely operates at its full power rating over a full year; the ratio between annual average power and installed capacity rating is the capacity factor. The installed capacity is the sum of all generator nameplate power ratings.

Ten of the largest hydroelectric producers as at 2014.				
Country	**Annual hydroelectric production (TWh)**	**Installed capacity (GW)**	**Capacity factor**	**% of total production**
China	1064	311	0.37	18.7%
Canada	383	76	0.59	58.3%
Brazil	373	89	0.56	63.2%
United States	282	102	0.42	6.5%
Russia	177	51	0.42	16.7%
India	132	40	0.43	10.2%
Norway	129	31	0.49	96.0%
Japan	87	50	0.37	8.4%
Venezuela	87	15	0.67	68.3%
France	69	25	0.46	12.2%

Major Projects Under Construction

Name	Maximum Capacity	Country	Construction started	Scheduled completion	Comments
Belo Monte Dam	11,181 MW	Brazil	March, 2011	2015	Preliminary construction underway. Construction suspended 14 days by court order Aug 2012
Siang Upper HE Project	11,000 MW	India	April, 2009	2024	Multi-phase construction over a period of 15 years. Construction was delayed due to dispute with China.
Tasang Dam	7,110 MW	Burma	March, 2007	2022	Controversial 228 meter tall dam with capacity to produce 35,446 GWh annually.
Xiangjiaba Dam	6,400 MW	China	November 26, 2006	2015	The last generator was commissioned on July 9, 2014
Grand Ethiopian Renaissance Dam	6,000 MW	Ethiopia	2011	2017	Located in the upper Nile Basin, drawing complaint from Egypt
Nuozhadu Dam	5,850 MW	China	2006	2017	

Name	Maximum Capacity	Country	Construction started	Scheduled completion	Comments
Jinping 2 Hydropower Station	4,800 MW	China	January 30, 2007	2014	To build this dam, 23 families and 129 local residents need to be moved. It works with Jinping 1 Hydropower Station as a group.
Diamer-Bhasha Dam	4,500 MW	Pakistan	October 18, 2011	2023	
Jinping 1 Hydropower Station	3,600 MW	China	November 11, 2005	2014	The sixth and final generator was commissioned on 15 July 2014
Jirau Power Station	3,300 MW	Brazil	2008	2013	Construction halted in March 2011 due to worker riots.
Guanyinyan Dam	3,000 MW	China	2008	2015	Construction of the roads and spillway started.
Lianghekou Dam	3,000 MW	China	2014	2023	
Dagangshan Dam	2,600 MW	China	August 15, 2008	2016	
Liyuan Dam	2,400 MW	China	2008	2013	
Tocoma Dam Bolívar State	2,160 MW	Venezuela	2004	2014	This power station would be the last development in the Low Caroni Basin, bringing the total to six power stations on the same river, including the 10,000MW Guri Dam.
Ludila Dam	2,100 MW	China	2007	2015	Brief construction halt in 2009 for environmental assessment.
Shuangjiangkou Dam	2,000 MW	China	December, 2007	2018	The dam will be 312 m high.
Ahai Dam	2,000 MW	China	July 27, 2006	2015	
Teles Pires Dam	1,820 MW	Brazil	2011	2015	
Site C Dam	1,100 MW	Canada	2015	2024	First large dam in western Canada since 1984
Lower Subansiri Dam	2,000 MW	India	2007	2016	

Pumped-storage Hydroelectricity

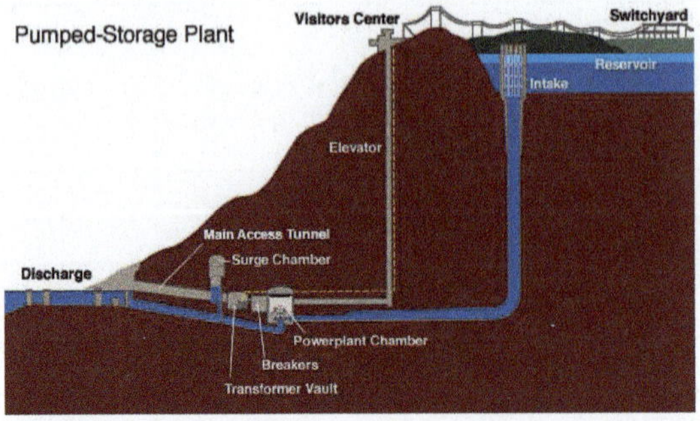

Diagram of the TVA pumped storage facility at Raccoon Mountain Pumped-Storage Plant

Pumped-storage hydroelectricity (PSH), or pumped hydroelectric energy storage (PHES), is a type of hydroelectric energy storage used by electric power systems for load balancing. The method stores energy in the form of gravitational potential energy of water, pumped from a lower elevation reservoir to a higher elevation. Low-cost surplus off-peak electric power is typically used to run the pumps. During periods of high electrical demand, the stored water is released through turbines to produce electric power. Although the losses of the pumping process makes the plant a net consumer of energy overall, the system increases revenue by selling more electricity during periods of *peak demand*, when electricity prices are highest.

Shaded-relief topo map of the Taum Sauk pumped storage plant in Missouri

Pumped-storage hydroelectricity allows energy from intermittent sources (such as solar, wind) and other renewables, or excess electricity from continuous base-load sources (such as coal or nuclear) to be saved for periods of higher demand. The reservoirs used with pumped storage are quite small when compared to conventional hydroelectric dams of similar power capacity, and generating periods are often less than half a day.

Pumped storage is the largest-capacity form of grid energy storage available, and, as of 2017, the DOE Global Energy Storage Database reports that PSH accounts for over 96% of all active tracked storage installations worldwide, with a total installed nameplate capacity of over 168 GW. The *round-trip* energy efficiency of PSH varies between 70%–80%, with some sources claiming up to 87%. The main disadvantage of PHS is the specialist nature of the site required, needing both geographical height and water availability. Suitable sites are therefore likely to be in hilly or mountainous regions, and potentially in areas of outstanding natural beauty, and therefore there are also social and ecological issues to overcome.

Overview

At times of low electrical demand, excess generation capacity is used to pump water into the upper reservoir. When there is higher demand, water is released back into the lower reservoir through a turbine, generating electricity. Reversible turbine/generator assemblies act as a combined pump and turbine generator unit (usually a Francis turbine design). In open-loop systems, pure pumped-storage plants store water in an upper reservoir with no natural inflows, while pump-back plants utilize a combination of pumped storage and conventional hydroelectric plants with an upper reservoir that is replenished in part by natural inflows from a stream or river. Plants that do not use pumped-storage are referred to as conventional hydroelectric plants; conventional hydroelectric plants that have significant storage capacity may be able to play a similar role in the electrical grid as pumped storage by deferring output until needed.

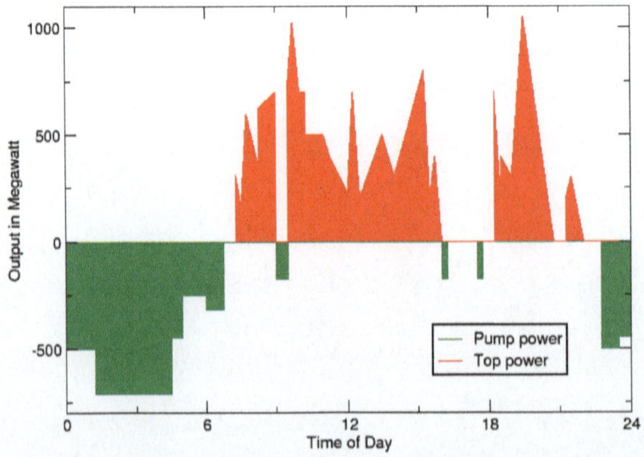

Power distribution, over a day, of a pumped-storage hydroelectricity facility. Green represents power consumed in pumping; red is power generated.

Taking into account evaporation losses from the exposed water surface and conversion losses, energy recovery of 70-80% or more can be regained. This technique is currently the most cost-effective means of storing large amounts of electrical energy, but capital costs and the presence of appropriate geography are critical decision factors in selecting pumped-storage plant sites.

The relatively low energy density of pumped storage systems requires either large flows and/or large differences in height between reservoirs. The only way to store a significant amount of energy is by having a large body of water located relatively near, but as high above as possible, a second body of water. In some places this occurs naturally, in others one or both bodies of water were man-made. Projects in which both reservoirs are artificial and in which no natural inflows are involved with either reservoir are referred to as "closed loop" systems.

These systems may be economical because they flatten out load variations on the power grid, permitting thermal power stations such as coal-fired plants and nuclear power plants that provide base-load electricity to continue operating at peak efficiency, while reducing the need for "peaking" power plants that use the same fuels as many base-load thermal plants, gas and oil, but have been designed for flexibility rather than maximal efficiency. Hence pumped storage systems are crucial when coordinating large groups of heterogeneous generators. However, capital costs for pumped-storage plants are relatively high.

Along with energy management, pumped storage systems help control electrical network frequency and provide reserve generation. Thermal plants are much less able to respond to sudden changes in electrical demand, potentially causing frequency and voltage instability. Pumped storage plants, like other hydroelectric plants, can respond to load changes within seconds.

The most important use for pumped storage has traditionally been to balance baseload powerplants, but may also be used to abate the fluctuating output of intermittent energy sources.

Pumped storage provides a load at times of high electricity output and low electricity demand, enabling additional system peak capacity. In certain jurisdictions, electricity prices may be close to zero or occasionally negative on occasions that there is more electrical generation available than there is load available to absorb it; although at present this is rarely due to wind or solar power alone, increased wind and solar generation will increase the likelihood of such occurrences. It is particularly likely that pumped storage will become especially important as a balance for very large scale photovoltaic generation. Increased long distance transmission capacity combined with significant amounts of energy storage will be a crucial part of regulating any large-scale deployment of intermittent renewable power sources. The high non-firm renewable electricity penetration in some regions supplies 40% of annual output, but 60% may be reached before additional storage is necessary.

There are small-scale installations of such technology, namely in buildings, although these are economically unfeasible given the economies of scale present. Also, a large volume of water is required for a meaningful storage capacity which is a difficult fit for an urban setting. Nevertheless, some authors defend its technological simplicity and secure provision of water as important externalities.

History

The first use of pumped storage was in the 1890s in Italy and Switzerland. In the 1930s reversible hydroelectric turbines became available. These turbines could operate as both turbine-generators and in reverse as electric motor driven pumps. The latest in large-scale engineering technology are variable speed machines for greater efficiency. These machines operate in synchronization with the network frequency when generating, but operate asynchronously (independent of the network frequency) when pumping.

The first use of pumped-storage in the United States was in 1930 by the Connecticut Electric and Power Company, using a large reservoir located near New Milford, Connecticut, pumping water from the Housatonic River to the storage reservoir 230 feet above.

Worldwide Use

In 2009, world pumped storage generating capacity was 104 GW, while other sources claim 127 GW, which comprises the vast majority of all types of utility grade electric storage. The EU had 38.3 GW net capacity (36.8% of world capacity) out of a total of 140 GW of hydropower and representing 5% of total net electrical capacity in the EU. Japan had 25.5 GW net capacity (24.5% of world capacity).

In 2010 the United States had 21.5 GW of pumped storage generating capacity (20.6% of world capacity). PHS generated (net) -5.501 GWh of energy in 2010 in the US because more energy is consumed in pumping than is generated. Nameplate pumped storage capacity had grown to 21.6 GW by 2014, with pumped storage comprising 97% of grid-scale energy storage in the US. As of late 2014, there were 51 active project proposals with a total of 39 GW of new nameplate capacity across all stages of the FERC licensing process for new pumped storage hydroelectric plants in the US, but no new plants were currently under construction in the US at the time.

The five largest operational pumped-storage plants are listed below

Station	Country	Location	Capacity (MW)
Bath County Pumped Storage Station	United States	38°12′32″N 79°48′00″W38.20889°N 79.80000°W	3,003
Guangdong Pumped Storage Power Station	China	23°45′52″N 113°57′12″E23.76444°N 113.95333°E	2,400
Huizhou Pumped Storage Power Station	China	23°16′07″N 114°18′50″E23.26861°N 114.31389°E	2,400
Okutataragi Pumped Storage Power Station	Japan	35°14′13″N 134°49′55″E35.23694°N 134.83194°E	1,932
Ludington Pumped Storage Power Plant	United States	43°53′37″N 86°26′43″W43.89361°N 86.44528°W	1,872

Pump-back Hydroelectric Dams

Conventional hydroelectric dams may also make use of pumped storage in a hybrid system that both generates power from water naturally flowing into the reservoir as well as storing water pumped back to the reservoir from below the dam. The Grand Coulee Dam in the US was expanded with a pump-back system in 1973. Existing dams may be repowered with reversing turbines thereby extending the length of time the plant can operate at capacity. Optionally a pump back powerhouse such as the Russell Dam(1992) may be added to a dam for increased generating capacity. Making use of an existing dams upper reservoir and transmission system can expedite projects and reduce costs.

Potential Technologies

Seawater

Pumped storage plants can operate on seawater, although there are additional challenges involved compared to operation with fresh water. In 1999, the 30 MW Yanbaru project in Okinawa was the first demonstration of seawater pumped storage. A 300 MW seawater-based project has recently been proposed on Lanai, Hawaii, and several seawater-based projects have been proposed in Ireland and Chile.

Direct Pumping

A new concept is to use wind turbines or solar power to drive water pumps directly. This could provide a more efficient process and usefully smooth out the variability of energy captured from the wind or sun. In northern Chile, the Espejo de Tarapacá project is a power project of Valhalla, that combines solar and hydroelectric resources. The project takes advantage of the unique geographic characteristics of the Atacama Desert in order to build a 300 MW pumped storage hydroelectric plant that uses the Pacific Ocean as its lower reservoir and an existing natural concavity as its upper reservoir, and a 600 MW-AC solar photovoltaic plant that is located in the region with the highest solar irradiation in the world. The above characteristics minimize the environmental impact and the cost of the plant, which is USD 400 MM for the pumped storage and USD 900 MM for the solar power plant.

Underground Reservoirs

The use of underground reservoirs has been investigated. Recent examples include the proposed Summit project in Norton, Ohio, the proposed Maysville project in Kentucky (underground limestone mine), and the Mount Hope project in New Jersey, which was to have used a former iron mine as the lower reservoir. Several new underground pumped storage projects have been proposed. Cost-per-kilowatt estimates for these projects can be lower than for surface projects if they use existing underground mine space. There are limited opportunities involving suitable underground space, but the number of underground pumped storage opportunities may increase if abandoned coal mines prove suitable.

Decentralised Systems

Small pumped-storage hydropower plants can be built on streams and within infrastructures, such as drinking water networks and artificial snow making infrastructures. Such plants provide distributed energy storage and distributed flexible electricity production and can contribute to the decentralized integration of intermittent renewable energy technologies, such as wind power and solar power. Reservoirs that can be used for small pumped-storage hydropower plants could include natural or artificial lakes, reservoirs within other structures such as irrigation, or unused portions of mines or underground military installations. In Switzerland one study suggested that the total installed capacity of small pumped-storage hydropower plants in 2011 could be increased by 3 to 9 times by providing adequate policy instruments.

Underwater Reservoirs

In March 2017 the research project StEnSea (Storing Energy at Sea) announced their successful completion of a four-week test of a pumped storage underwater reservoir. In this configuration a hollow sphere submerged in deep water acts as the lower reservoir while the upper reservoir is the enclosing body of water. When a reversible turbine integrated into the sphere uses surplus electricity to pump water out of the sphere the force of the pump must act on the entire column of water above the sphere, so the deeper the sphere is located, the more potential energy it can store and convert back to electricity by letting water back in via the turbine.

As such the energy storage capacity of the submerged reservoir is not governed by the gravitational energy in the traditional sense, but rather by the vertical pressure variation.

While StEnSea's test took place at a depth of 100 m in the fresh water Lake Constance, the technology is foreseen to be used in salt water at greater depths. Since the submerged reservoir needs only a connecting electrical cable, the depth at which it can be employed is limited only by the depth at which the turbine can function, currently limited to 700 m. The challenge of designing salt water pumped storage in this underwater configuration brings a range of advantages:

- No land area is required,
- No mechanical structure other than the electrical cable needs to span the distance of the potential energy difference,
- In the presence of sufficient seabed area multiple reservoirs can scale the storage capacity

without limits,

- Should a reservoir collapse, the consequences would be limited apart from the loss of the reservoir itself,
- Evaporation from the upper reservoir has no effect on the energy conversion efficiency,
- Transmission of electricity between the reservoir and the grid can be established from a nearby offshore wind farm limiting transmission loss and obviating the need for onshore cabling permits.

A current commercial design featuring a sphere with an inner diameter of 30 m submerged to 700 m would correspond to a 20 MWh capacity which with a 5 MW turbine would lead to a 4-hour discharge time. An energy park with multiple such reservoirs would bring the storage cost to around a few eurocents per kWh with construction and equipment costs in the range € 1200-1400 € per kW. To avoid excessive transmission cost and loss, the reservoirs should be placed off deep water coasts of densely populated areas, such as Norway, Spain, USA and Japan. With this limitation the concept would allow for worldwide electricity storage of close to 900 GWh.

For comparison, a traditional, gravity-based pumped storage capable of storing 20 MWh in a water reservoir the size of a 30 m sphere would need a hydraulic head of 519 m with the elevation spanned by a pressurized water pipe requiring typically a hill or mountain for support.

Run-of-the-river Hydroelectricity

Chief Joseph Dam near Bridgeport, Washington, USA, is a major run-of-the-river station without a sizeable reservoir.

Run-of-the-river hydroelectricity (ROR) or run-of-river hydroelectricity is a type of hydroelectric generation plant whereby little or no water storage is provided. Run-of-the-river power plants may have no water storage at all or a limited amount of storage, in which case the storage reservoir is referred to as pondage. A plant without pondage has no water storage and is, therefore, subject to seasonal river flows. Thus, the plant will operate as an intermittent energy source while a plant with pondage can regulate the water flow at all times and can serve as a peaking power plant or base load power plant.

Concept

Run-of-the-river or ROR hydroelectricity is considered ideal for streams or rivers that can sustain a minimum flow or those regulated by a lake or reservoir upstream. A small dam is usually built to create a headpond ensuring that there is enough water entering the penstock pipes that lead to the turbines which are at a lower elevation. Projects with pondage, as opposed to those without pondage, can store water for daily load demands. In general, projects divert some or most of a river's flow (up to 95% of mean annual discharge) through a pipe and/or tunnel leading to electricity-generating turbines, then return the water back to the river downstream.

ROR projects are dramatically different in design and appearance from conventional hydroelectric projects. Traditional hydro dams store enormous quantities of water in reservoirs, sometimes flooding large tracts of land. In contrast, run-of-river projects do not have most of the disadvantages associated with dams and reservoirs, which is why they are often considered environmentally friendly.

Saint Marys Falls - run of the river (1902)

The use of the term "run-of-the-river" for power projects varies around the world. Some may consider a project ROR if power is produced with no water storage while limited storage is considered ROR by others. Developers may mislabel a project ROR to soothe public perception about its environmental or social effects. The Bureau of Indian Standards describes run-of-the-river hydroelectricity as:

A power station utilizing the run of the river flows for generation of power with sufficient pondage for supplying water for meeting diurnal or weekly fluctuations of demand. In such stations, the normal course of the river is not materially altered.

Many of the larger ROR projects have been designed to a scale and generating capacity rivaling some traditional hydro dams. For example, the Beauharnois Hydroelectric Generating Station in Quebec is rated at 1,853 MW and a 2006 proposal in British Columbia, Canada has been designed to generate 1027 megawatts capacity. Some run of the river projects are downstream of other dams and reservoirs. The run of the river project didn't build the reservoir, but does take advantage of the water supplied by it. An example would be the 1995, 1,436 MW La Grande-1 generating station, previous upstream dams and reservoirs are part of the 1980s James Bay Project.

Advantages

When developed with care to footprint size and location, ROR hydro projects can create sustainable energy minimizing impacts to the surrounding environment and nearby communities. Advantages include:

Cleaner Power, Fewer Greenhouse Gases

Like all hydro-electric power, run-of-the-river hydro harnesses the natural potential energy of water, eliminating the need to burn coal or natural gas to generate the electricity needed by consumers and industry. Moreover, run-of-the-river hydro-electric plants do not require the formation of a reservoir (unlike conventional hydro-electric plants) thus eliminating the methane and carbon dioxide emissions generally caused by the decomposition of organic matter in the reservoir of a hydro-electric dam.

Less Flooding/Reservoirs

Substantial flooding of the upper part of the river is not required for run-of-river projects as a large reservoir is not required. As a result, people living at or near the river don't need to be relocated and natural habitats and productive farmlands are not wiped out.

Disadvantages

"Unfirm" Power

Run-of-the-River power is considered an "unfirm" source of power: a run-of-the-river project has little or no capacity for energy storage and hence can't co-ordinate the output of electricity generation to match consumer demand. It thus generates much more power during times when seasonal river flows are high (i.e., spring freshet), and depending on location, much less during drier summer months or frozen winter months.

Availability of Sites

The potential power at a site is a result of the head and flow of water. By damming a river, the head is available to generate power at the face of the dam. Where a dam may create a reservoir hundreds of kilometers long, in run of the river the head is usually delivered by a canal, pipe or tunnel constructed upstream of the power house. Due to the cost of upstream construction, a steep drop in the river is desirable.

Environmental Impacts

Small, well-sited ROR projects can be developed with minimal environmental impacts. Larger projects have more environmental concerns. For example, Plutonic Power Corp.'s canceled Bute Inlet Hydroelectric Project in BC would have seen three clusters of run-of-river projects with 17 river diversions; as proposed, this run-of-river project would divert over 90 kilometres of streams and rivers into tunnels and pipelines, requiring 443 km of new transmission line, 267 km of permanent roads, and 142 bridges to be built in wilderness areas.

British Columbia's mountainous terrain and wealth of big rivers have made it a global testing ground for run-of-river technology. As of March 2010, there were 628 applications pending for new water licences solely for the purposes of power generation – representing more than 750 potential points of river diversion.

Concerns

- Diverting large amounts of river water reduces river flows, affecting water velocity and depth, minimizing habitat quality for fish and aquatic organisms; reduced flows can lead to excessively warm water for salmon and other fish in summer.
- New access roads and transmission lines can cause extensive habitat fragmentation for many species, allowing the introduction of invasive species and increases in undesirable human activities, like illegal hunting.
- The lack of reservoir storage may result in intermittent operation, reducing the project's viability.

Major Examples

- Belo Monte Dam, 11,233 megawatts (15,064,000 hp), Pará, Brazil
- Chief Joseph Dam, 2,620 megawatts (3,510,000 hp)
- Beauharnois Hydroelectric Power Station, 1,903 megawatts (2,552,000 hp)
- Bonneville Dam, 1,092 megawatts (1,464,000 hp)
- Satluj Jal Vidyut Nigam Ltd, Satluj River, Shimla, India, 1,500 megawatts (2,000,000 hp)
- Ghazi-Barotha Hydropower Project on River Indus in Pakistan, 1,450 megawatts (1,940,000 hp)
- La Grande-1 generating station, 1,436 megawatts (1,926,000 hp)
- Kohala Hydropower Project, Jhelum River, Muzaffarabad, Pakistan, 1,100 megawatts (1,500,000 hp)
- Carillon Generating Station, Quebec, Canada, 752 megawatts (1,008,000 hp)
- Upper Tamakoshi Project, Nepal, 456 MW

References

- "Turbines Off NYC East River Will Create Enough Energy to Power 9,500 Homes". U.S. Department of Energy. Retrieved 13 February 2012
- Bent Sørensen (2004). Renewable Energy: Its Physics, Engineering, Use, Environmental Impacts, Economy, and Planning Aspects. Academic Press. pp. 556–. ISBN 978-0-12-656153-1
- De Oliveira e Silva, Guilherme; Hendrick, Patrick (2016-10-01). "Pumped hydro energy storage in buildings". Applied Energy. 179: 1242–1250. doi:10.1016/j.apenergy.2016.07.046
- Şentürk, Fuat (1994). Hydraulics of dams and reservoirs (reference. ed.). Highlands Ranch, Colo.: Water Resources Publications. p. 375. ISBN 0-918334-80-2
- "Norske oppfinneres turbinteknologi kan bli brukt i britisk tidevannseventyr". Teknisk Ukeblad. 14 January 2017. Retrieved 15 January 2017
- Turcotte, D. L.; Schubert, G. (2002). "Chapter 4". Geodynamics (2nd ed.). Cambridge, England, UK: Cambridge University Press. pp. 136–137. ISBN 978-0-521-66624-4

- Oliver, Antony (9 June 2015). "Swansea Tidal Lagoon power plant wins planning permission". Infrastructure Intelligence. Retrieved 6 July 2015
- Minchinton, W. E. (October 1979). "Early Tide Mills: Some Problems". Technology and Culture. Society for the History of Technology. 20 (4): 777–786. JSTOR 3103639. doi:10.2307/3103639

2

Wind Power and Wave Power

Wind power is the conversion of airflow into electric power. Wind power is a renewable source of energy that does not produce any greenhouse gases. The topics discussed in the chapter are of great importance to broaden the existing knowledge on wind and wave power.

Wind Power

Wind power stations in Xinjiang, China

Wind power is the use of air flow through wind turbines to mechanically power generators for electric power. Wind power, as an alternative to burning fossil fuels, is plentiful, renewable, widely distributed, clean, produces no greenhouse gas emissions during operation, consumes no water, and uses little land. The net effects on the environment are far less problematic than those of non-renewable power sources.

Wind farms consist of many individual wind turbines which are connected to the electric power transmission network. Onshore wind is an inexpensive source of electric power, competitive with or in many places cheaper than coal or gas plants. Offshore wind is steadier and stronger than on land, and offshore farms have less visual impact, but construction and maintenance costs are considerably higher. Small onshore wind farms can feed some energy into the grid or provide electric power to isolated off-grid locations.

Wind power gives variable power which is very consistent from year to year but which has significant variation over shorter time scales. It is therefore used in conjunction with other electric power sources to give a reliable supply. As the proportion of wind power in a region increases, a need to upgrade the grid, and a lowered ability to supplant conventional production can occur. Power management techniques such as having excess capacity, geographically distributed turbines, dispatchable backing sources, sufficient hydroelectric power, exporting and importing power to neighboring areas, using vehicle-to-grid strategies or reducing demand when wind production is low, can in many cases overcome these problems. In addition, weather forecasting permits the electric power network to be readied for the predictable variations in production that occur.

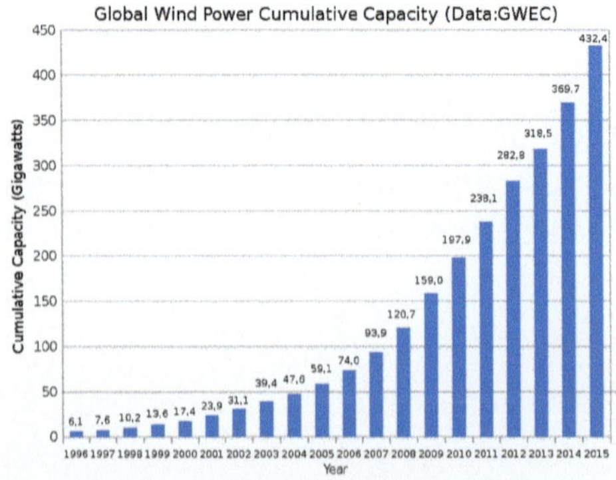

Global growth of installed capacity

As of 2015, Denmark generates 40% of its electric power from wind, and at least 83 other countries around the world are using wind power to supply their electric power grids. In 2014 global wind power capacity expanded 16% to 369,553 MW. Yearly wind energy production is also growing rapidly and has reached around 4% of worldwide electric power usage, 11.4% in the EU.

History

Charles Brush's windmill of 1888, used for generating electric power.

Wind power has been used as long as humans have put sails into the wind. For more than two millennia wind-powered machines have ground grain and pumped water. Wind power was widely available and not confined to the banks of fast-flowing streams, or later, requiring sources of fuel. Wind-powered pumps drained the polders of the Netherlands, and in arid regions such as the American mid-west or the Australian outback, wind pumps provided water for live stock and steam engines.

The first windmill used for the production of electric power was built in Scotland in July 1887 by Prof James Blyth of Anderson's College, Glasgow (the precursor of Strathclyde University). Blyth's

10 metres (33 ft) high, cloth-sailed wind turbine was installed in the garden of his holiday cottage at Marykirk in Kincardineshire and was used to charge accumulators developed by the Frenchman Camille Alphonse Faure, to power the lighting in the cottage, thus making it the first house in the world to have its electric power supplied by wind power. Blyth offered the surplus electric power to the people of Marykirk for lighting the main street, however, they turned down the offer as they thought electric power was "the work of the devil." Although he later built a wind turbine to supply emergency power to the local Lunatic Asylum, Infirmary and Dispensary of Montrose the invention never really caught on as the technology was not considered to be economically viable.

Across the Atlantic, in Cleveland, Ohio a larger and heavily engineered machine was designed and constructed in the winter of 1887–1888 by Charles F. Brush, this was built by his engineering company at his home and operated from 1886 until 1900. The Brush wind turbine had a rotor 17 metres (56 ft) in diameter and was mounted on an 18 metres (59 ft) tower. Although large by today's standards, the machine was only rated at 12 kW. The connected dynamo was used either to charge a bank of batteries or to operate up to 100 incandescent light bulbs, three arc lamps, and various motors in Brush's laboratory.

With the development of electric power, wind power found new applications in lighting buildings remote from centrally-generated power. Throughout the 20th century parallel paths developed small wind stations suitable for farms or residences, and larger utility-scale wind generators that could be connected to electric power grids for remote use of power. Today wind powered generators operate in every size range between tiny stations for battery charging at isolated residences, up to near-gigawatt sized offshore wind farms that provide electric power to national electrical networks.

Wind Farms

Large onshore wind farms		
Wind farm	**Current capacity (MW)**	**Country**
Gansu Wind Farm	6,000	China
Muppandal wind farm	1,500	India
Alta (Oak Creek-Mojave)	1,320	United States
Jaisalmer Wind Park	1,064	India
Shepherds Flat Wind Farm	845	United States
Roscoe Wind Farm	782	United States
Horse Hollow Wind Energy Center	736	United States
Capricorn Ridge Wind Farm	662	United States
Fântânele-Cogealac Wind Farm	600	Romania
Fowler Ridge Wind Farm	600	United States
Whitelee Wind Farm	539	United Kingdom

A wind farm is a group of wind turbines in the same location used for production of electric power. A large wind farm may consist of several hundred individual wind turbines distributed over an extended area, but the land between the turbines may be used for agricultural or other purposes. For example, Gansu Wind Farm, the largest wind farm in the world, has several thousand turbines. A wind farm may also be located offshore.

Almost all large wind turbines have the same design — a horizontal axis wind turbine having an upwind rotor with three blades, attached to a nacelle on top of a tall tubular tower.

In a wind farm, individual turbines are interconnected with a medium voltage (often 34.5 kV), power collection system and communications network. In general, a distance of 7D (7 × Rotor Diameter of the Wind Turbine) is set between each turbine in a fully developed wind farm. At a substation, this medium-voltage electric current is increased in voltage with a transformer for connection to the high voltage electric power transmission system.

Generator Characteristics and Stability

Induction generators, which were often used for wind power projects in the 1980s and 1990s, require reactive power for excitation so substations used in wind-power collection systems include substantial capacitor banks for power factor correction. Different types of wind turbine generators behave differently during transmission grid disturbances, so extensive modelling of the dynamic electromechanical characteristics of a new wind farm is required by transmission system operators to ensure predictable stable behaviour during system faults. In particular, induction generators cannot support the system voltage during faults, unlike steam or hydro turbine-driven synchronous generators.

Today these generators aren't used any more in modern turbines. Instead today most turbines use variable speed generators combined with partial- or full-scale power converter between the turbine generator and the collector system, which generally have more desirable properties for grid interconnection and have Low voltage ride through-capabilities. Modern concepts use either doubly fed machines with partial-scale converters or squirrel-cage induction generators or synchronous generators (both permanently and electrically excited) with full scale converters.

Transmission systems operators will supply a wind farm developer with a grid code to specify the requirements for interconnection to the transmission grid. This will include power factor, constancy of frequency and dynamic behaviour of the wind farm turbines during a system fault.

Offshore Wind Power

Offshore wind power refers to the construction of wind farms in large bodies of water to generate electric power. These installations can utilize the more frequent and powerful winds that are available in these locations and have less aesthetic impact on the landscape than land based projects. However, the construction and the maintenance costs are considerably higher.

Siemens and Vestas are the leading turbine suppliers for offshore wind power. DONG Energy, Vattenfall and E.ON are the leading offshore operators. As of October 2010, 3.16 GW of offshore wind power capacity was operational, mainly in Northern Europe. According to BTM Consult, more than 16 GW of additional capacity will be installed before the end of 2014 and the UK and Germany will become the two leading markets. Offshore wind power capacity is expected to reach a total of 75 GW worldwide by 2020, with significant contributions from China and the US.

The world's second full-scale floating wind turbine (and first to be installed without the use of heavy-lift vessels), WindFloat, operating at rated capacity (2 MW) approximately 5 km offshore of Póvoa de Varzim, Portugal

In 2012, 1,662 turbines at 55 offshore wind farms in 10 European countries produced 18 TWh, enough to power almost five million households. As of August 2013 the London Array in the United Kingdom is the largest offshore wind farm in the world at 630 MW. This is followed by Gwynt y Môr (576 MW), also in the UK.

World's largest offshore wind farms				
Wind farm	**Capacity (MW)**	**Country**	**Turbines and model**	**Commissioned**
London Array	630	United Kingdom	175 × Siemens SWT-3.6	2012
Gwynt y Môr	576	United Kingdom	160 × Siemens SWT-3.6 107	2015
Greater Gabbard	504	United Kingdom	140 × Siemens SWT-3.6	2012
Anholt	400	Denmark	111 × Siemens SWT-3.6–120	2013
BARD Offshore 1	400	Germany	80 BARD 5.0 turbines	2013

Collection and Transmission Network

In a wind farm, individual turbines are interconnected with a medium voltage (usually 34.5 kV) power collection system and communications network. At a substation, this medium-voltage electric current is increased in voltage with a transformer for connection to the high voltage electric power transmission system.

A transmission line is required to bring the generated power to (often remote) markets. For an off-shore station this may require a submarine cable. Construction of a new high-voltage line may be too costly for the wind resource alone, but wind sites may take advantage of lines installed for conventionally fueled generation.

One of the biggest current challenges to wind power grid integration in the United States is the necessity of developing new transmission lines to carry power from wind farms, usually in remote lowly populated states in the middle of the country due to availability of wind, to high load locations, usually on the coasts where population density is higher. The current transmission lines in remote locations were not designed for the transport of large amounts of energy. As transmission lines become longer the losses associated with power transmission increase, as modes of losses at lower lengths are exacerbated and new modes of losses are no longer negligible as the length is increased, making it harder to transport large loads over large distances. However, resistance from state and local governments makes it difficult to construct new transmission lines. Multi state power transmission projects are discouraged by states with cheap electric power rates for fear that exporting their cheap power will lead to increased rates. A 2005 energy law gave the Energy Department authority to approve transmission projects states refused to act on, but after an attempt to use this authority, the Senate declared the department was being overly aggressive in doing so. Another problem is that wind companies find out after the fact that the transmission capacity of a new farm is below the generation capacity, largely because federal utility rules to encourage renewable energy installation allow feeder lines to meet only minimum standards. These are important issues that need to be solved, as when the transmission capacity does not meet the generation capacity, wind farms are forced to produce below their full potential or stop running all together, in a process known as curtailment. While this leads to potential renewable generation left untapped, it prevents possible grid overload or risk to reliable service.

Wind Power Capacity and Production

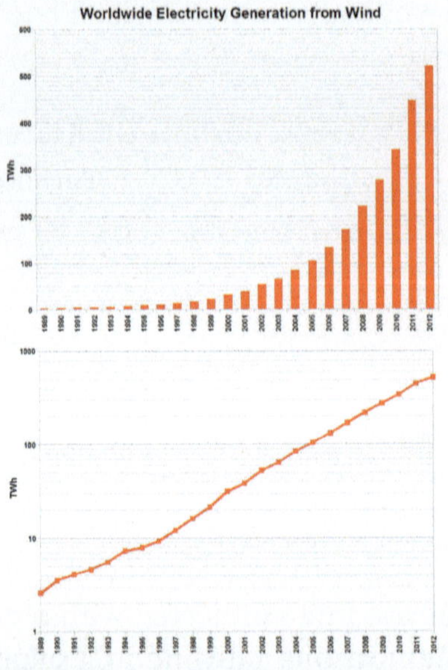

Worldwide wind generation up to 2012

Worldwide there are now over two hundred thousand wind turbines operating, with a total nameplate capacity of 432 GW as of end 2015. The European Union alone passed some 100 GW nameplate capacity in September 2012, while the United States surpassed 75 GW in 2015 and China's grid connected capacity passed 145 GW in 2015.

World wind generation capacity more than quadrupled between 2000 and 2006, doubling about every three years. The United States pioneered wind farms and led the world in installed capacity in the 1980s and into the 1990s. In 1997 installed capacity in Germany surpassed the U.S. and led until once again overtaken by the U.S. in 2008. China has been rapidly expanding its wind installations in the late 2000s and passed the U.S. in 2010 to become the world leader. As of 2011, 83 countries around the world were using wind power on a commercial basis.

Wind power capacity has expanded rapidly to 336 GW in June 2014, and wind energy production was around 4% of total worldwide electric power usage, and growing rapidly. The actual amount of electric power that wind is able to generate is calculated by multiplying the nameplate capacity by the capacity factor, which varies according to equipment and location. Estimates of the capacity factors for wind installations are in the range of 35% to 44%. By doubling installed capacity an average 40% increase in electricity production would be expected.

Europe accounted for 48% of the world total wind power generation capacity in 2009. In 2010, Spain became Europe's leading producer of wind energy, achieving 42,976 GWh. Germany held the top spot in Europe in terms of installed capacity, with a total of 27,215 MW as of 31 December 2010. In 2015 wind power constituted 15.6% of all installed power generation capacity in the EU and it generates around 11.4% of its power.

Top wind-power electric power producing countries in 2012		
Country	**Wind-power Production (TWh)**	**% of World Total**
United States	140.9	26.4%
China	118.1	22.1%
Spain	49.1	9.2%
Germany	46.0	8.6%
India	30.0	5.6%
United Kingdom	19.6	3.7%
France	14.9	2.8%
Italy	13.4	2.5%
Canada	11.8	2.2%
Denmark	10.3	1.9%
(rest of world)	80.2	15.0%
World Total	**534.3**	**100.0%**

Growth Trends

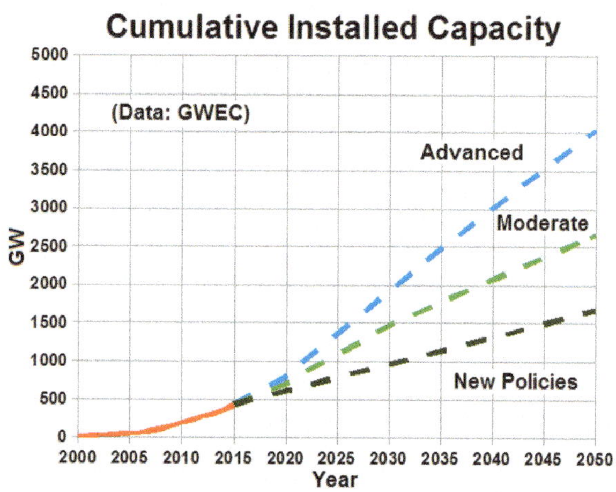

Worldwide installed wind power capacity forecast

The wind power industry set new records in 2014 – more than 50 GW of new capacity was installed. Another record breaking year occurred in 2015, with 22% annual market growth resulting in the 60 GW mark being passed. In 2015, close to half of all new wind power was added outside of the traditional markets in Europe and North America. This was largely from new construction in China and India. Global Wind Energy Council (GWEC) figures show that 2015 recorded an increase of installed capacity of more than 63 GW, taking the total installed wind energy capacity to 432.9 GW, up from 74 GW in 2006. In terms of economic value, the wind energy sector has become one of the important players in the energy markets, with the total investments reaching US$329bn (€296.6bn), an increase of 4% over 2014.

Although the wind power industry was affected by the global financial crisis in 2009 and 2010, GWEC predicts that the installed capacity of wind power will be 792.1 GW by the end of 2020 and 4,042 GW by end of 2050. The increased commissioning of wind power is being accompanied by record low prices for forthcoming renewable electric power. In some cases, wind onshore is already the cheapest electric power generation option and costs are continuing to decline. The contracted prices for wind onshore for the next few years are now as low as 30 USD/MWh.

In the EU in 2015, 44% of all new generating capacity was wind power; while in the same period net fossil fuel power capacity decreased.

Capacity Factor

Since wind speed is not constant, a wind farm's annual energy production is never as much as the sum of the generator nameplate ratings multiplied by the total hours in a year. The ratio of actual productivity in a year to this theoretical maximum is called the capacity factor. Typical capacity factors are 15–50%; values at the upper end of the range are achieved in favourable sites and are due to wind turbine design improvements.

Online data is available for some locations, and the capacity factor can be calculated from the yearly output. For example, the German nationwide average wind power capacity factor over all of 2012 was just under 17.5% (45,867 GW·h/yr / (29.9 GW × 24 × 366) = 0.1746), and the capacity factor for Scottish wind farms averaged 24% between 2008 and 2010.

Unlike fueled generating plants, the capacity factor is affected by several parameters, including the variability of the wind at the site and the size of the generator relative to the turbine's swept area. A small generator would be cheaper and achieve a higher capacity factor but would produce less electric power (and thus less profit) in high winds. Conversely, a large generator would cost more but generate little extra power and, depending on the type, may stall out at low wind speed. Thus an optimum capacity factor of around 40–50% would be aimed for.

A 2008 study released by the U.S. Department of Energy noted that the capacity factor of new wind installations was increasing as the technology improves, and projected further improvements for future capacity factors. In 2010, the department estimated the capacity factor of new wind turbines in 2010 to be 45%. The annual average capacity factor for wind generation in the US has varied between 29.8% and 34.0% during the period 2010–2015.

Penetration

Country	Penetration
Denmark (2015)	42.1%
Portugal (2013)	23%
Spain (2011)	16%
Ireland (2012)	16%
United Kingdom (2015)	11%
Germany (2011)	8%
United States (2016)	5.5%

Wind energy penetration is the fraction of energy produced by wind compared with the total generation. The wind power penetration in world electric power generation in 2015 was 3.5%.

There is no generally accepted maximum level of wind penetration. The limit for a particular grid will depend on the existing generating plants, pricing mechanisms, capacity for energy storage, demand management and other factors. An interconnected electric power grid will already include reserve generating and transmission capacity to allow for equipment failures. This reserve capacity can also serve to compensate for the varying power generation produced by wind stations. Studies have indicated that 20% of the total annual electrical energy consumption may be incorporated with minimal difficulty. These studies have been for locations with geographically dispersed wind farms, some degree of dispatchable energy or hydropower with storage capacity, demand management, and interconnected to a large grid area enabling the export of electric power when needed. Beyond the 20% level, there are few technical limits, but the economic implications become more significant. Electrical utilities continue to study the effects of large scale penetration of wind generation on system stability and economics.

A wind energy penetration figure can be specified for different duration of time, but is often quoted annually. To obtain 100% from wind annually requires substantial long term storage or substantial interconnection to other systems which may already have substantial storage. On a monthly, weekly, daily, or hourly basis—or less—wind might supply as much as or more than 100% of current use, with the rest stored or exported. Seasonal industry might then take advantage of high wind and low usage times such as at night when wind output can exceed normal demand. Such industry might include production of silicon, aluminum, steel, or of natural gas, and hydrogen, and using future long term storage to facilitate 100% energy from variable renewable energy. Homes can also be programmed to accept extra electric power on demand, for example by remotely turning up water heater thermostats.

In Australia, the state of South Australia generates around half of the nation's wind power capacity. By the end of 2011 wind power in South Australia, championed by Premier (and Climate Change Minister) Mike Rann, reached 26% of the State's electric power generation, edging out coal for the first time. At this stage South Australia, with only 7.2% of Australia's population, had 54% of Australia's installed capacity.

Variability

Wind turbines are typically installed in favorable windy locations. In the image, wind power generators in Spain, near an Osborne bull.

Electric power generated from wind power can be highly variable at several different timescales: hourly, daily, or seasonally. Annual variation also exists, but is not as significant. Because instantaneous electrical generation and consumption must remain in balance to maintain grid stability, this variability can present substantial challenges to incorporating large amounts of wind power into a grid system. Intermittency and the non-dispatchable nature of wind energy production can raise costs for regulation, incremental operating reserve, and (at high penetration levels) could require an increase in the already existing energy demand management, load shedding, storage solutions or system interconnection with HVDC cables. The variability of wind is quite different from solar, wind may be producing power at night when other baseload plants are often overproducing.

Fluctuations in load and allowance for failure of large fossil-fuel generating units requires operating reserve capacity, which can be increased to compensate for variability of wind generation.

Wind power is variable, and during low wind periods it must be replaced by other power sources. Transmission networks presently cope with outages of other generation plants and daily changes in electrical demand, but the variability of intermittent power sources such as wind power, is more frequent than those of conventional power generation plants which, when scheduled to be operating, may be able to deliver their nameplate capacity around 95% of the time.

Presently, grid systems with large wind penetration require a small increase in the frequency of usage of natural gas spinning reserve power plants to prevent a loss of electric power in the event that there is no wind. At low wind power penetration, this is less of an issue.

GE has installed a prototype wind turbine with onboard battery similar to that of an electric car, equivalent of 1 minute of production. Despite the small capacity, it is enough to guarantee that power output complies with forecast for 15 minutes, as the battery is used to eliminate the difference rather than provide full output. In certain cases the increased predictability can be used to take wind power penetration from 20 to 30 or 40 per cent. The battery cost can be retrieved by selling burst power on demand and reducing backup needs from gas plants.

In the UK there were 124 separate occasions from 2008 to 2010 when the nation's wind output fell to less than 2% of installed capacity. A report on Denmark's wind power noted that their wind power network provided less than 1% of average demand on 54 days during the year 2002. Wind power advocates argue that these periods of low wind can be dealt with by simply restarting existing power stations that have been held in readiness, or interlinking with HVDC. Electrical grids with slow-responding thermal power plants and without ties to networks with hydroelectric generation may have to limit the use of wind power. According to a 2007 Stanford University study published in the *Journal of Applied Meteorology and Climatology*, interconnecting ten or more wind farms can allow an average of 33% of the total energy produced (i.e. about 8% of total nameplate capacity) to be used as reliable, baseload electric power which can be relied on to handle peak loads, as long as minimum criteria are met for wind speed and turbine height.

Conversely, on particularly windy days, even with penetration levels of 16%, wind power generation can surpass all other electric power sources in a country. In Spain, in the early hours of 16 April 2012 wind power production reached the highest percentage of electric power production till then, at 60.46% of the total demand. In Denmark, which had power market penetration of 30% in 2013, over 90 hours, wind power generated 100% of the country's power, peaking at 122% of the country's demand at 2 am on 28 October.

Increase in system operation costs, Euros per MWh, for 10% & 20% wind share		
Country	**10%**	**20%**
Germany	2.5	3.2
Denmark	0.4	0.8
Finland	0.3	1.5
Norway	0.1	0.3
Sweden	0.3	0.7

A 2006 International Energy Agency forum presented costs for managing intermittency as a function of wind-energy's share of total capacity for several countries, as shown in the table on the right. Three reports on the wind variability in the UK issued in 2009, generally agree that variability of wind needs to be taken into account by adding 20% to the operating reserve, but it does not make the grid unmanageable. The additional costs, which are modest, can be quantified.

The combination of diversifying variable renewables by type and location, forecasting their variation, and integrating them with dispatchable renewables, flexible fueled generators, and demand response can create a power system that has the potential to meet power supply needs reliably. Integrating ever-higher levels of renewables is being successfully demonstrated in the real world:

In 2009, eight American and three European authorities, writing in the leading electrical engineers' professional journal, didn't find "a credible and firm technical limit to the amount of wind energy that can be accommodated by electric power grids". In fact, not one of more than 200 international studies, nor official studies for the eastern and western U.S. regions, nor the International Energy Agency, has found major costs or technical barriers to reliably integrating up to 30% variable renewable supplies into the grid, and in some studies much more.

Solar power tends to be complementary to wind. On daily to weekly timescales, high pressure areas tend to bring clear skies and low surface winds, whereas low pressure areas tend to be windier and cloudier. On seasonal timescales, solar energy peaks in summer, whereas in many areas wind energy is lower in summer and higher in winter. Thus the seasonal variation of wind and solar power tend to cancel each other somewhat. In 2007 the Institute for Solar Energy Supply Technology of the University of Kassel pilot-tested a combined power plant linking solar, wind, biogas and hydrostorage to provide load-following power around the clock and throughout the year, entirely from renewable sources.

Predictability

Wind power forecasting methods are used, but predictability of any particular wind farm is low for short-term operation. For any particular generator there is an 80% chance that wind output will change less than 10% in an hour and a 40% chance that it will change 10% or more in 5 hours.

However, studies by Graham Sinden (2009) suggest that, in practice, the variations in thousands of wind turbines, spread out over several different sites and wind regimes, are smoothed. As the distance between sites increases, the correlation between wind speeds measured at those sites, decreases.

Thus, while the output from a single turbine can vary greatly and rapidly as local wind speeds vary, as more turbines are connected over larger and larger areas the average power output becomes less variable and more predictable.

Wind power hardly ever suffers major technical failures, since failures of individual wind turbines have hardly any effect on overall power, so that the distributed wind power is reliable and predictable, whereas conventional generators, while far less variable, can suffer major unpredictable outages.

Energy Storage

The Sir Adam Beck Generating Complex at Niagara Falls, Canada, includes a large pumped-storage hydroelectricity reservoir. During hours of low electrical demand excess electrical grid power is used to pump water up into the reservoir, which then provides an extra 174 MW of electric power during periods of peak demand.

Typically, conventional hydroelectricity complements wind power very well. When the wind is blowing strongly, nearby hydroelectric stations can temporarily hold back their water. When the wind drops they can, provided they have the generation capacity, rapidly increase production to compensate. This gives a very even overall power supply and virtually no loss of energy and uses no more water.

Alternatively, where a suitable head of water is not available, pumped-storage hydroelectricity or other forms of grid energy storage such as compressed air energy storage and thermal energy storage can store energy developed by high-wind periods and release it when needed. The type of storage needed depends on the wind penetration level – low penetration requires daily storage, and high penetration requires both short and long term storage – as long as a month or more. Stored energy increases the economic value of wind energy since it can be shifted to displace higher cost generation during peak demand periods. The potential revenue from this arbitrage can offset the cost and losses of storage. For example, in the UK, the 1.7 GW Dinorwig pumped-storage plant evens out electrical demand peaks, and allows base-load suppliers to run their plants more efficiently. Although pumped-storage power systems are only about 75% efficient, and have high installation costs, their low running costs and ability to reduce the required electrical base-load can save both fuel and total electrical generation costs.

In particular geographic regions, peak wind speeds may not coincide with peak demand for electrical power. In the U.S. states of California and Texas, for example, hot days in summer may have low wind speed and high electrical demand due to the use of air conditioning. Some utilities subsidize the purchase of geothermal heat pumps by their customers, to reduce electric power demand during the summer months by making air conditioning up to 70% more efficient; widespread adoption of this technology would better match electric power demand to wind availability in areas with hot summers and low summer winds. A possible future option may be to interconnect widely dispersed geographic areas with an HVDC “super grid”. In the U.S. it is estimated that to upgrade the transmission system to take in planned or potential renewables would cost at least USD 60 bn, while the society value of added windpower would be more than that cost.

Germany has an installed capacity of wind and solar that can exceed daily demand, and has been exporting peak power to neighboring countries, with exports which amounted to some 14.7 billion kWh in 2012. A more practical solution is the installation of thirty days storage capacity able to supply 80% of demand, which will become necessary when most of Europe’s energy is obtained from wind power and solar power. Just as the EU requires member countries to maintain 90 days strategic reserves of oil it can be expected that countries will provide electric power storage, instead of expecting to use their neighbors for net metering.

Capacity Credit, Fuel Savings and Energy Payback

The capacity credit of wind is estimated by determining the capacity of conventional plants displaced by wind power, whilst maintaining the same degree of system security. According to the American Wind Energy Association, production of wind power in the United States in 2015 avoided consumption of 73 billion gallons of water and reduced CO_2 emissions by 132 million metric tons, while providing USD 7.3 bn in public health savings.

The energy needed to build a wind farm divided into the total output over its life, Energy Return on Energy Invested, of wind power varies but averages about 20–25. Thus, the energy payback time is typically around one year.

Economics

Wind turbines reached grid parity (the point at which the cost of wind power matches traditional sources) in some areas of Europe in the mid-2000s, and in the US around the same time. Falling prices continue to drive the levelized cost down and it has been suggested that it has reached general grid parity in Europe in 2010, and will reach the same point in the US around 2016 due to an expected reduction in capital costs of about 12%.

Electric Power Cost and Trends

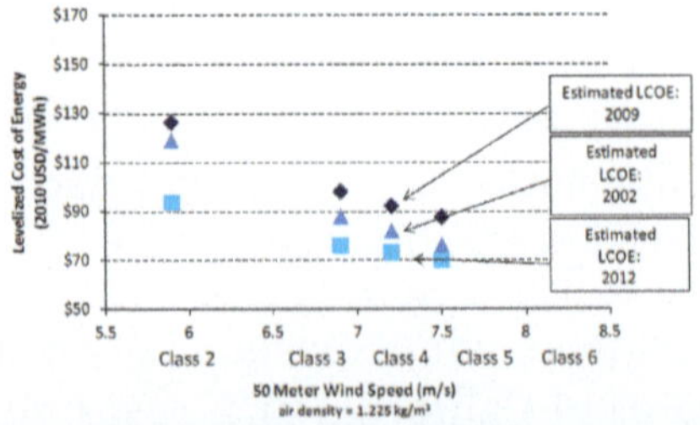

Estimated cost per MWh for wind power in Denmark

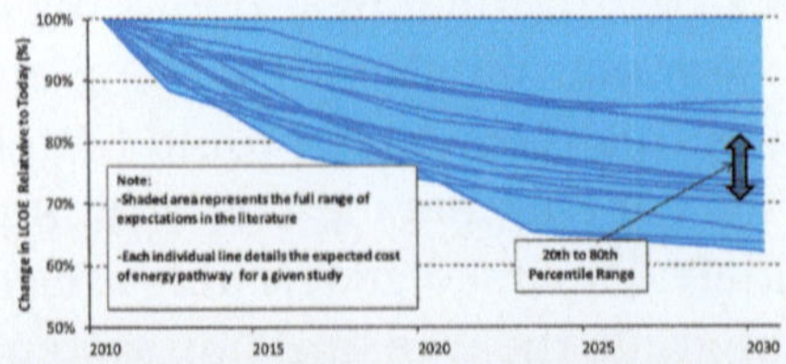

The National Renewable Energy Laboratory projects that the levelized cost of wind power in the U.S. will decline about 25% from 2012 to 2030.

A turbine blade convoy passing through Edenfield in the U.K. (2008). Even longer two-piece blades are now manufactured, and then assembled on-site to reduce difficulties in transportation.

Wind power is capital intensive, but has no fuel costs. The price of wind power is therefore much more stable than the volatile prices of fossil fuel sources. The marginal cost of wind energy once a station is constructed is usually less than 1-cent per kW·h.

However, the estimated average cost per unit of electric power must incorporate the cost of construction of the turbine and transmission facilities, borrowed funds, return to investors (including cost of risk), estimated annual production, and other components, averaged over the projected useful life of the equipment, which may be in excess of twenty years. Energy cost estimates are highly dependent on these assumptions so published cost figures can differ substantially. In 2004, wind energy cost a fifth of what it did in the 1980s, and some expected that downward trend to continue as larger multi-megawatt turbines were mass-produced. In 2012 capital costs for wind turbines were substantially lower than 2008–2010 but still above 2002 levels. A 2011 report from the

American Wind Energy Association stated, "Wind's costs have dropped over the past two years, in the range of 5 to 6 cents per kilowatt-hour recently.... about 2 cents cheaper than coal-fired electric power, and more projects were financed through debt arrangements than tax equity structures last year.... winning more mainstream acceptance from Wall Street's banks.... Equipment makers can also deliver products in the same year that they are ordered instead of waiting up to three years as was the case in previous cycles.... 5,600 MW of new installed capacity is under construction in the United States, more than double the number at this point in 2010. Thirty-five percent of all new power generation built in the United States since 2005 has come from wind, more than new gas and coal plants combined, as power providers are increasingly enticed to wind as a convenient hedge against unpredictable commodity price moves."

A British Wind Energy Association report gives an average generation cost of onshore wind power of around 3.2 pence (between US 5 and 6 cents) per kW·h (2005). Cost per unit of energy produced was estimated in 2006 to be 5 to 6 percent above the cost of new generating capacity in the US for coal and natural gas: wind cost was estimated at $55.80 per MW·h, coal at $53.10/MW·h and natural gas at $52.50. Similar comparative results with natural gas were obtained in a governmental study in the UK in 2011. In 2011 power from wind turbines could be already cheaper than fossil or nuclear plants; it is also expected that wind power will be the cheapest form of energy generation in the future. The presence of wind energy, even when subsidised, can reduce costs for consumers (€5 billion/yr in Germany) by reducing the marginal price, by minimising the use of expensive peaking power plants.

An 2012 EU study shows base cost of onshore wind power similar to coal, when subsidies and externalities are disregarded. Wind power has some of the lowest external costs.

In February 2013 Bloomberg New Energy Finance (BNEF) reported that the cost of generating electric power from new wind farms is cheaper than new coal or new baseload gas plants. When including the current Australian federal government carbon pricing scheme their modeling gives costs (in Australian dollars) of $80/MWh for new wind farms, $143/MWh for new coal plants and $116/MWh for new baseload gas plants. The modeling also shows that "even without a carbon price (the most efficient way to reduce economy-wide emissions) wind energy is 14% cheaper than new coal and 18% cheaper than new gas." Part of the higher costs for new coal plants is due to high financial lending costs because of "the reputational damage of emissions-intensive investments". The expense of gas fired plants is partly due to "export market" effects on local prices. Costs of production from coal fired plants built in "the 1970s and 1980s" are cheaper than renewable energy sources because of depreciation. In 2015 BNEF calculated LCOE prices per MWh energy in new powerplants (excluding carbon costs) : $85 for onshore wind ($175 for offshore), $66–75 for coal in the Americas ($82–105 in Europe), gas $80–100. A 2014 study showed unsubsidized LCOE costs between $37–81, depending on region. A 2014 US DOE report showed that in some cases power purchase agreement prices for wind power had dropped to record lows of $23.5/MWh.

The cost has reduced as wind turbine technology has improved. There are now longer and lighter wind turbine blades, improvements in turbine performance and increased power generation efficiency. Also, wind project capital and maintenance costs have continued to decline. For example, the wind industry in the USA in early 2014 were able to produce more power at lower cost by using taller wind turbines with longer blades, capturing the faster winds at higher elevations. This has opened up new opportunities and in Indiana, Michigan, and Ohio, the price of power from wind turbines built 300 feet to 400 feet above the ground can now compete with conventional fossil fuels like coal. Prices have fallen to about 4 cents per kilowatt-hour in some cases and utilities have been increasing the amount of wind energy in their portfolio, saying it is their cheapest option.

A number of initiatives are working to reduce costs of electric power from offshore wind. One example is the Carbon Trust Offshore Wind Accelerator, a joint industry project, involving nine offshore wind developers, which aims to reduce the cost of offshore wind by 10% by 2015. It has been suggested that innovation at scale could deliver 25% cost reduction in offshore wind by 2020. Henrik Stiesdal, former Chief Technical Officer at Siemens Wind Power, has stated that by 2025 energy from offshore wind will be one of the cheapest, scalable solutions in the UK, compared to other renewables and fossil fuel energy sources, if the true cost to society is factored into the cost of energy equation. He calculates the cost at that time to be 43 EUR/MWh for onshore, and 72 EUR/MWh for offshore wind.

Incentives and Community Benefits

U.S. landowners typically receive \$3,000–\$5,000 annual rental income per wind turbine, while farmers continue to grow crops or graze cattle up to the foot of the turbines. Shown: the Brazos Wind Farm, Texas.

Some of the 6,000 turbines in California's Altamont Pass Wind Farm aided by tax incentives during the 1980s.

The U.S. wind industry generates tens of thousands of jobs and billions of dollars of economic activity. Wind projects provide local taxes, or payments in lieu of taxes and strengthen the economy of rural communities by providing income to farmers with wind turbines on their land. Wind energy in many jurisdictions receives financial or other support to encourage its development. Wind energy benefits from subsidies in many jurisdictions, either to increase its attractiveness, or to compensate for subsidies received by other forms of production which have significant negative externalities.

In the US, wind power receives a production tax credit (PTC) of 1.5¢/kWh in 1993 dollars for each kW·h produced, for the first ten years; at 2.2 cents per kW·h in 2012, the credit was renewed on 2 January 2012, to include construction begun in 2013. A 30% tax credit can be applied instead of receiving the PTC. Another tax benefit is accelerated depreciation. Many American states also provide incentives, such as exemption from property tax, mandated purchases, and additional markets for "green credits". The Energy Improvement and Extension Act of 2008 contains extensions of credits for wind, including microturbines. Countries such as Canada and Germany also provide incentives for wind turbine construction, such as tax credits or minimum purchase prices for wind generation, with assured grid access (sometimes referred to as feed-in tariffs). These feed-in tariffs are typically set well above average electric power prices. In December 2013 U.S. Senator Lamar Alexander and other Republican senators argued that the "wind energy production tax credit should be allowed to expire at the end of 2013" and it expired 1 January 2014 for new installations.

Secondary market forces also provide incentives for businesses to use wind-generated power, even if there is a premium price for the electricity. For example, socially responsible manufacturers pay utility companies a premium that goes to subsidize and build new wind power infrastructure. Companies use wind-generated power, and in return they can claim that they are undertaking strong "green" efforts. In the US the organization Green-e monitors business compliance with these renewable energy credits.

Small-scale Wind Power

A small Quietrevolution QR5 Gorlov type vertical axis wind turbine on the roof of Colston Hall in Bristol, England. Measuring 3 m in diameter and 5 m high, it has a nameplate rating of 6.5 kW.

Small-scale wind power is the name given to wind generation systems with the capacity to produce up to 50 kW of electrical power. Isolated communities, that may otherwise rely on diesel generators, may use wind turbines as an alternative. Individuals may purchase these systems to reduce or eliminate their dependence on grid electric power for economic reasons, or to reduce their carbon footprint. Wind turbines have been used for household electric power generation in conjunction with battery storage over many decades in remote areas.

Recent examples of small-scale wind power projects in an urban setting can be found in New York City, where, since 2009, a number of building projects have capped their roofs with Gorlov-type helical wind turbines. Although the energy they generate is small compared to the buildings' overall consumption, they help to reinforce the building's 'green' credentials in ways that "showing people your high-tech boiler" can not, with some of the projects also receiving the direct support of the New York State Energy Research and Development Authority.

Grid-connected domestic wind turbines may use grid energy storage, thus replacing purchased electric power with locally produced power when available. The surplus power produced by domestic microgenerators can, in some jurisdictions, be fed into the network and sold to the utility company, producing a retail credit for the microgenerators' owners to offset their energy costs.

Off-grid system users can either adapt to intermittent power or use batteries, photovoltaic or diesel systems to supplement the wind turbine. Equipment such as parking meters, traffic warning signs, street lighting, or wireless Internet gateways may be powered by a small wind turbine, possibly combined with a photovoltaic system, that charges a small battery replacing the need for a connection to the power grid.

A Carbon Trust study into the potential of small-scale wind energy in the UK, published in 2010, found that small wind turbines could provide up to 1.5 terawatt hours (TW·h) per year of electric power (0.4% of total UK electric power consumption), saving 0.6 million tonnes of carbon dioxide (Mt CO_2) emission savings. This is based on the assumption that 10% of households would install turbines at costs competitive with grid electric power, around 12 pence (US 19 cents) a kW·h. A report prepared for the UK's government-sponsored Energy Saving Trust in 2006, found that home power generators of various kinds could provide 30 to 40% of the country's electric power needs by 2050.

Distributed generation from renewable resources is increasing as a consequence of the increased awareness of climate change. The electronic interfaces required to connect renewable generation units with the utility system can include additional functions, such as the active filtering to enhance the power quality.

Environmental Effects

The environmental impact of wind power when compared to the environmental impacts of fossil fuels, is relatively minor. According to the IPCC, in assessments of the life-cycle global warming potential of energy sources, wind turbines have a median value of between 12 and 11 (gCO_2eq/kWh) depending on whether off- or onshore turbines are being assessed. Compared with other low carbon power sources, wind turbines have some of the lowest global warming potential per unit of electrical energy generated.

Livestock grazing near a wind turbine.

While a wind farm may cover a large area of land, many land uses such as agriculture are compatible with it, as only small areas of turbine foundations and infrastructure are made unavailable for use.

There are reports of bird and bat mortality at wind turbines as there are around other artificial structures. The scale of the ecological impact may or may not be significant, depending on specific circumstances. Prevention and mitigation of wildlife fatalities, and protection of peat bogs, affect the siting and operation of wind turbines.

Wind turbines generate some noise. At a residential distance of 300 metres (980 ft) this may be around 45 dB, which is slightly louder than a refrigerator. At 1.5 km (1 mi) distance they become inaudible. There are anecdotal reports of negative health effects from noise on people who live very close to wind turbines. Peer-reviewed research has generally not supported these claims.

The United States Air Force and Navy have expressed concern that siting large wind turbines near bases "will negatively impact radar to the point that air traffic controllers will lose the location of aircraft."

Aesthetic aspects of wind turbines and resulting changes of the visual landscape are significant. Conflicts arise especially in scenic and heritage protected landscapes.

Politics

Central Government

Nuclear power and fossil fuels are subsidized by many governments, and wind power and other forms of renewable energy are also often subsidized. For example, a 2009 study by the Environmental Law Institute assessed the size and structure of U.S. energy subsidies over the 2002–2008 period. The study estimated that subsidies to fossil-fuel based sources amounted to approximately $72 billion over this period and subsidies to renewable fuel sources totalled $29 billion. In the United States, the federal government has paid US$74 billion for energy subsidies to support R&D for nuclear power ($50 billion) and fossil fuels ($24 billion) from 1973 to 2003. During this same time frame, renewable energy technologies and energy efficiency received a total of US$26 billion.

It has been suggested that a subsidy shift would help to level the playing field and support growing energy sectors, namely solar power, wind power, and biofuels. History shows that no energy sector was developed without subsidies.

Part of the Seto Hill Windfarm in Japan.

According to the International Energy Agency (IEA) (2011), energy subsidies artificially lower the price of energy paid by consumers, raise the price received by producers or lower the cost of production. "Fossil fuels subsidies costs generally outweigh the benefits. Subsidies to renewables and low-carbon energy technologies can bring long-term economic and environmental benefits". In November 2011, an IEA report entitled *Deploying Renewables 2011* said "subsidies in green energy technologies that were not yet competitive are justified in order to give an incentive to investing into technologies with clear environmental and energy security benefits". The IEA's report disagreed with claims that renewable energy technologies are only viable through costly subsidies and not able to produce energy reliably to meet demand.

In the U.S., the wind power industry has recently increased its lobbying efforts considerably, spending about $5 million in 2009 after years of relative obscurity in Washington. By comparison, the U.S. nuclear industry alone spent over $650 million on its lobbying efforts and campaign contributions during a single ten-year period ending in 2008.

Following the 2011 Japanese nuclear accidents, Germany's federal government is working on a new plan for increasing energy efficiency and renewable energy commercialization, with a particular focus on offshore wind farms. Under the plan, large wind turbines will be erected far away from the coastlines, where the wind blows more consistently than it does on land, and where the enormous turbines won't bother the inhabitants. The plan aims to decrease Germany's dependence on energy derived from coal and nuclear power plants.

Public Opinion

Surveys of public attitudes across Europe and in many other countries show strong public support for wind power. About 80% of EU citizens support wind power. In Germany, where wind power has gained very high social acceptance, hundreds of thousands of people have invested in citizens' wind farms across the country and thousands of small and medium-sized enterprises are running successful businesses in a new sector that in 2008 employed 90,000 people and generated 8% of

Germany's electric power.

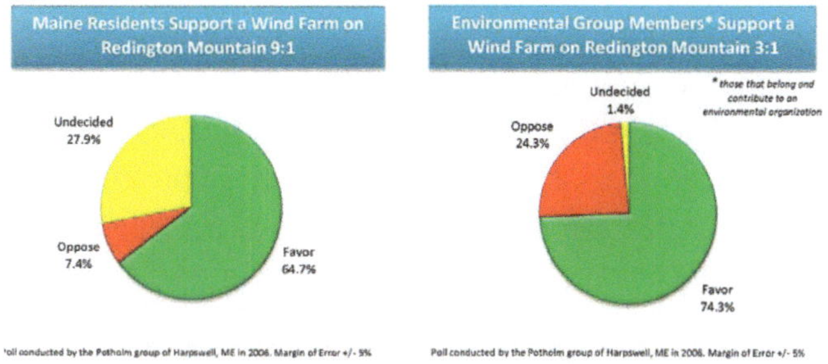

Environmental group members are both more in favor of wind power (74%) as well as more opposed (24%). Few are undecided.

Although wind power is a popular form of energy generation, the construction of wind farms is not universally welcomed, often for aesthetic reasons.

In Spain, with some exceptions, there has been little opposition to the installation of inland wind parks. However, the projects to build offshore parks have been more controversial. In particular, the proposal of building the biggest offshore wind power production facility in the world in south-western Spain in the coast of Cádiz, on the spot of the 1805 Battle of Trafalgar has been met with strong opposition who fear for tourism and fisheries in the area, and because the area is a war grave.

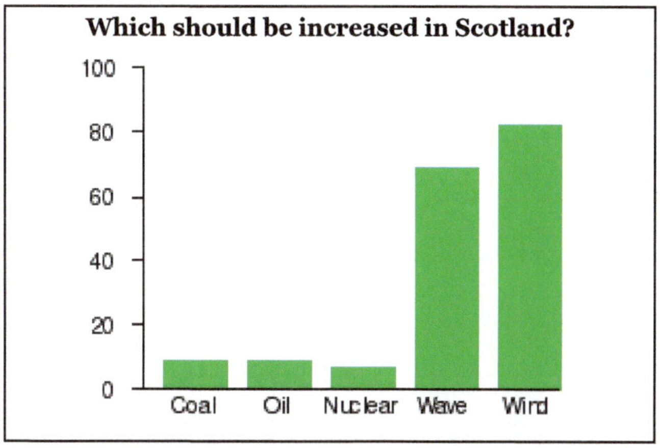

In a survey conducted by Angus Reid Strategies in October 2007, 89 per cent of respondents said that using renewable energy sources like wind or solar power was positive for Canada, because these sources were better for the environment. Only 4 per cent considered using renewable sources as negative since they can be unreliable and expensive. According to a Saint Consulting survey in April 2007, wind power was the alternative energy source most likely to gain public support for future development in Canada, with only 16% opposed to this type of energy. By contrast, 3 out of 4 Canadians opposed nuclear power developments.

A 2003 survey of residents living around Scotland's 10 existing wind farms found high levels of community acceptance and strong support for wind power, with much support from those who lived closest to the wind farms. The results of this survey support those of an earlier Scottish Executive survey 'Public attitudes to the Environment in Scotland 2002', which found that the Scottish public would prefer the majority of their electric power to come from renewables, and which rated wind power as the cleanest source of renewable energy. A survey conducted in 2005 showed that 74% of people in Scotland agree that wind farms are necessary to meet current and future energy needs. When people were asked the same question in a Scottish renewables study conducted in 2010, 78% agreed. The increase is significant as there were twice as many wind farms in 2010 as there were in 2005. The 2010 survey also showed that 52% disagreed with the statement that wind farms are "ugly and a blot on the landscape". 59% agreed that wind farms were necessary and that how they looked was unimportant. Regarding tourism, query responders consider power pylons, cell phone towers, quarries and plantations more negatively than wind farms. Scotland is planning to obtain 100% of electric power from renewable sources by 2020.

In other cases there is direct community ownership of wind farm projects. The hundreds of thousands of people who have become involved in Germany's small and medium-sized wind farms demonstrate such support there.

This 2010 Harris Poll reflects the strong support for wind power in Germany, other European countries, and the U.S.

Opinion on increase in number of wind farms, 2010 Harris Poll

	U.S.	Great Britain	France	Italy	Spain	Germany
	%	%	%	%	%	%
Strongly oppose	3	6	6	2	2	4
Oppose more than favour	9	12	16	11	9	14
Favour more than oppose	37	44	44	38	37	42
Strongly favour	50	38	33	49	53	40

Community

Wind turbines such as these, in Cumbria, England, have been opposed for a number of reasons, including aesthetics, by some sectors of the population.

Many wind power companies work with local communities to reduce environmental and other concerns associated with particular wind farms. In other cases there is direct community ownership of wind farm projects. Appropriate government consultation, planning and approval procedures also help to minimize environmental risks. Some may still object to wind farms but, according to The Australia Institute, their concerns should be weighed against the need to address the threats posed by climate change and the opinions of the broader community.

In America, wind projects are reported to boost local tax bases, helping to pay for schools, roads and hospitals. Wind projects also revitalize the economy of rural communities by providing steady income to farmers and other landowners.

In the UK, both the National Trust and the Campaign to Protect Rural England have expressed concerns about the effects on the rural landscape caused by inappropriately sited wind turbines and wind farms.

A panoramic view of the United Kingdom's Whitelee Wind Farm with Lochgoin Reservoir in the foreground.

Some wind farms have become tourist attractions. The Whitelee Wind Farm Visitor Centre has an exhibition room, a learning hub, a café with a viewing deck and also a shop. It is run by the Glasgow Science Centre.

In Denmark, a loss-of-value scheme gives people the right to claim compensation for loss of value of their property if it is caused by proximity to a wind turbine. The loss must be at least 1% of the property's value.

Despite this general support for the concept of wind power in the public at large, local opposition often exists and has delayed or aborted a number of projects. For example, there are concerns that some installations can negatively affect TV and radio reception and Doppler weather radar, as well as produce excessive sound and vibration levels leading to a decrease in property values. Potential broadcast-reception solutions include predictive interference modeling as a component of site selection. A study of 50,000 home sales near wind turbines found no statistical evidence that prices were affected.

While aesthetic issues are subjective and some find wind farms pleasant and optimistic, or symbols of energy independence and local prosperity, protest groups are often formed to attempt to block new wind power sites for various reasons.

This type of opposition is often described as NIMBYism, but research carried out in 2009 found that there is little evidence to support the belief that residents only object to renewable power facilities such as wind turbines as a result of a "Not in my Back Yard" attitude.

Turbine Design

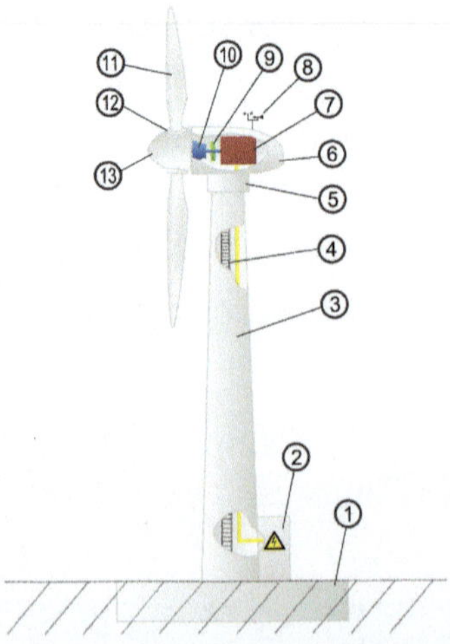

Typical wind turbine components:

1. Foundation
2. Connection to the electric grid Tower
3. Access ladder
4. Wind orientation control (Yaw control)
5. Nacelle
6. Generator
7. Anemometer
8. Electric or Mechanical Brake
9. Gearbox
10. Gearbox
11. Rotor blade
12. Blade pitch control
13. Rotor hub

Typical components of a wind turbine (gearbox, rotor shaft and brake assembly) being lifted into position

Wind turbines are devices that convert the wind's kinetic energy into electrical power. The result of over a millennium of windmill development and modern engineering, today's wind turbines are manufactured in a wide range of horizontal axis and vertical axis types. The smallest turbines are used for applications such as battery charging for auxiliary power. Slightly larger turbines can be used for making small contributions to a domestic power supply while selling unused power back to the utility supplier via the electrical grid. Arrays of large turbines, known as wind farms, have become an increasingly important source of renewable energy and are used in many countries as part of a strategy to reduce their reliance on fossil fuels.

Wind turbine design is the process of defining the form and specifications of a wind turbine to extract energy from the wind. A wind turbine installation consists of the necessary systems needed to capture the wind's energy, point the turbine into the wind, convert mechanical rotation into electrical power, and other systems to start, stop, and control the turbine.

In 1919 the German physicist Albert Betz showed that for a hypothetical ideal wind-energy extraction machine, the fundamental laws of conservation of mass and energy allowed no more than 16/27 (59.3%) of the kinetic energy of the wind to be captured. This Betz limit can be approached in modern turbine designs, which may reach 70 to 80% of the theoretical Betz limit.

The aerodynamics of a wind turbine are not straightforward. The air flow at the blades is not the same as the airflow far away from the turbine. The very nature of the way in which energy is extracted from the air also causes air to be deflected by the turbine. In addition the aerodynamics of a wind turbine at the rotor surface exhibit phenomena that are rarely seen in other aerodynamic fields. The shape and dimensions of the blades of the wind turbine are determined by the aerodynamic performance required to efficiently extract energy from the wind, and by the strength required to resist the forces on the blade.

In addition to the aerodynamic design of the blades, the design of a complete wind power system must also address the design of the installation's rotor hub, nacelle, tower structure, generator, controls, and foundation. Turbine design makes extensive use of computer modelling and simulation tools. These are becoming increasingly sophisticated as highlighted by a recent state-of-the-art review by Hewitt et al. Further design factors must also be considered when integrating wind turbines into electrical power grids.

Wind Energy

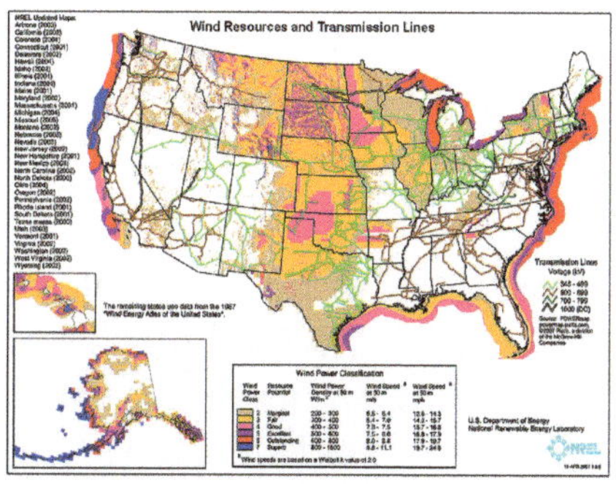

Map of available wind power for the United States. Color codes indicate wind power density class.

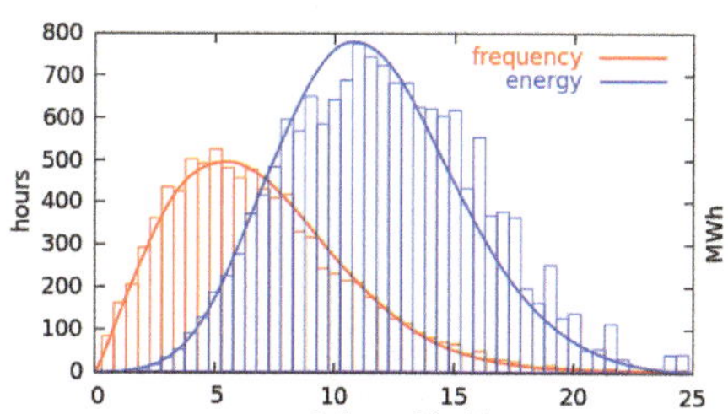

Distribution of wind speed (red) and energy (blue) for all of 2002 at the Lee Ranch facility in Colorado. The histogram shows measured data, while the curve is the Rayleigh model distribution for the same average wind speed.

Wind energy is the kinetic energy of air in motion, also called wind. Total wind energy flowing through an imaginary surface with area A during the time t is:

$$E = \frac{1}{2}mv^2 = \frac{1}{2}(Avt\rho)v^2 = \frac{1}{2}At\rho v^3,$$

where ρ is the density of air; v is the wind speed; Avt is the volume of air passing through A (which is considered perpendicular to the direction of the wind); $Avt\rho$ is therefore the mass m passing through "A". Note that $\frac{1}{2}\rho v^2$ is the kinetic energy of the moving air per unit volume.

Power is energy per unit time, so the wind power incident on A (e.g. equal to the rotor area of a wind turbine) is:

$$P = \frac{E}{t} = \frac{1}{2}A\rho v^3.$$

Wind power in an open air stream is thus *proportional* to the *third power* of the wind speed; the available power increases eightfold when the wind speed doubles. Wind turbines for grid electric power therefore need to be especially efficient at greater wind speeds.

Wind is the movement of air across the surface of the Earth, affected by areas of high pressure and of low pressure. The global wind kinetic energy averaged approximately 1.50 MJ/m² over the period from 1979 to 2010, 1.31 MJ/m² in the Northern Hemisphere with 1.70 MJ/m² in the Southern Hemisphere. The atmosphere acts as a thermal engine, absorbing heat at higher temperatures, releasing heat at lower temperatures. The process is responsible for production of wind kinetic energy at a rate of 2.46 W/m² sustaining thus the circulation of the atmosphere against frictional dissipation. A global 1 km² map of wind resources is housed based on calculations by the Technical University of Denmark. Unlike 'static' wind speed atlases which give a single average speed across multiple years, tools such as Renewables.ninja provide time-varying simulations of wind speed and power output from different wind turbine models at hourly resolution.

The total amount of economically extractable power available from the wind is considerably more than present human power use from all sources. Axel Kleidon of the Max Planck Institute in Germany, carried out a "top down" calculation on how much wind energy there is, starting with the incoming solar radiation that drives the winds by creating temperature differences in the atmosphere. He concluded that somewhere between 18 TW and 68 TW could be extracted.

Cristina Archer and Mark Z. Jacobson presented a "bottom-up" estimate, which unlike Kleidon's are based on actual measurements of wind speeds, and found that there is 1700 TW of wind power at an altitude of 100 metres over land and sea. Of this, "between 72 and 170 TW could be extracted in a practical and cost-competitive manner". They later estimated 80 TW. However research at Harvard University estimates 1 Watt/m² on average and 2–10 MW/km² capacity for large scale wind farms, suggesting that these estimates of total global wind resources are too high by a factor of about 4.

The strength of wind varies, and an average value for a given location does not alone indicate the amount of energy a wind turbine could produce there.

To assess prospective wind power sites a probability distribution function is often fit to the observed wind speed data. Different locations will have different wind speed distributions. The Weibull model closely mirrors the actual distribution of hourly/ten-minute wind speeds at many locations. The Weibull factor is often close to 2 and therefore a Rayleigh distribution can be used as a less accurate, but simpler model.

Windmill

The smock mill Goliath in front of the wind farm Growind in Eemshaven in the Netherlands

A windmill is a mill that converts the energy of wind into rotational energy by means of vanes called sails or blades. Centuries ago, windmills usually were used to mill grain (gristmills), pump water (windpumps), or both. The majority of modern windmills take the form of wind turbines used to generate electricity, or windpumps used to pump water, either for land drainage or to extract groundwater.

Windmills in Antiquity

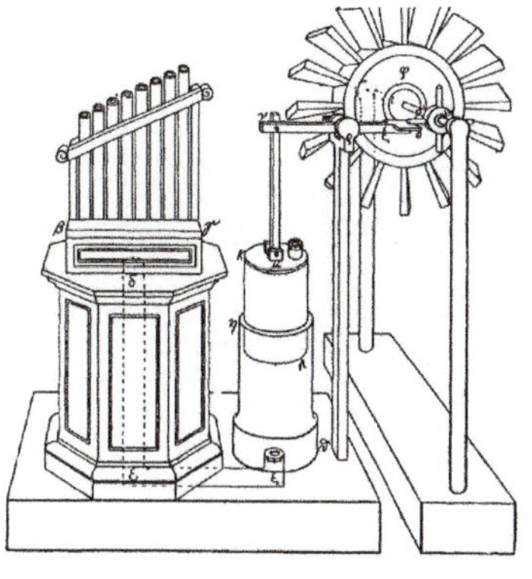

Heron's wind-powered organ

The windwheel of the Greek engineer Heron of Alexandria in the first century is the earliest known instance of using a wind-driven wheel to power a machine. Another early example of a wind-driven wheel was the prayer wheel, which has been used in Tibet and China since the fourth century.

It has been claimed that the Babylonian emperor Hammurabi planned to use wind power for his ambitious irrigation project in the seventeenth century BCE.

Horizontal Windmills

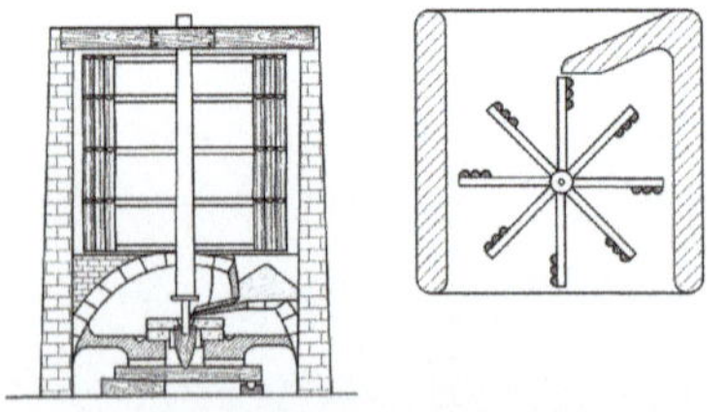

The Persian horizontal windmill

Hooper's Mill, Margate, Kent, an eighteenth-century European horizontal windmill

Coat of arms of Jemgum, East Friesland, Germany.

The first practical windmills had sails that rotated in a horizontal plane, around a vertical axis. According to Ahmad Y. al-Hassan, these panemone windmills were invented in eastern Persia as recorded by the Persian geographer Estakhri in the ninth century. The authenticity of an earlier anecdote of a windmill involving the second caliph Umar (AD 634–644) is questioned on the grounds that it appears in a tenth-century document. Made of six to 12 sails covered in reed matting or cloth material, these windmills were used to grind grain or draw up water, and were quite different from the later European vertical windmills. Windmills were in widespread use across the Middle East and Central Asia, and later spread to China and India from there.

A similar type of horizontal windmill with rectangular blades, used for irrigation, can also be found in thirteenth-century China (during the Jurchen Jin dynasty in the north), introduced by the travels of Yelü Chucai to Turkestan in 1219.

Horizontal windmills were built, in small numbers, in Europe during the 18th and nineteenth centuries, for example Fowler's Mill at Battersea in London, and Hooper's Mill at Margate in Kent. These early modern examples seem not to have been directly influenced by the horizontal windmills of the Middle and Far East, but to have been independent inventions by engineers influenced by the Industrial Revolution.

Vertical Windmills

Due to a lack of evidence, debate occurs among historians as to whether or not Middle Eastern horizontal windmills triggered the original development of European windmills. In northwestern Europe, the horizontal-axis or vertical windmill (so called due to the plane of the movement of its sails) is believed to date from the last quarter of the twelfth century in the triangle of northern France, eastern England and Flanders.

The earliest certain reference to a windmill in Europe (assumed to have been of the vertical type) dates from 1185, in the former village of Weedley in Yorkshire which was located at the southern tip of the Wold overlooking the Humber estuary. A number of earlier, but less certainly dated, twelfth-century European sources referring to windmills have also been found. These earliest mills were used to grind cereals.

Post Mill

The evidence at present is that the earliest type of European windmill was the post mill, so named because of the large upright post on which the mill's main structure (the "body" or "buck") is balanced. By mounting the body this way, the mill is able to rotate to face the wind direction; an essential requirement for windmills to operate economically in north-western Europe, where wind directions are variable. The body contains all the milling machinery. The first post mills were of the sunken type, where the post was buried in an earth mound to support it. Later, a wooden support was developed called the trestle. This was often covered over or surrounded by a roundhouse to protect the trestle from the weather and to provide storage space. This type of windmill was the most common in Europe until the nineteenth century, when more powerful tower and smock mills replaced them.

Hollow-post Mill

In a hollow-post mill, the post on which the body is mounted is hollowed out, to accommodate the drive shaft. This makes it possible to drive machinery below or outside the body while still being able to rotate the body into the wind. Hollow-post mills driving scoop wheels were used in the Netherlands to drain wetlands from the fourteenth century onwards.

Tower Mill

Tower mills in Spain

By the end of the thirteenth century, the masonry tower mill, on which only the cap is rotated rather than the whole body of the mill, had been introduced. The spread of tower mills came with a growing economy that called for larger and more stable sources of power, though they were more expensive to build. In contrast to the post mill, only the cap of the tower mill needs to be turned into the wind, so the main structure can be made much taller, allowing the sails to be made longer, which enables them to provide useful work even in low winds. The cap can be turned into the wind either by winches or gearing inside the cap or from a winch on the tail pole outside the mill. A method of keeping the cap and sails into the wind automatically is by using a fantail, a small windmill mounted at right angles to the sails, at the rear of the windmill. These are also fitted to tail poles of post mills and are common in Great Britain and English-speaking countries of the former British Empire, Denmark, and Germany but rare in other places. Around some parts of the Mediterranean Sea, tower mills with fixed caps were built because the wind's direction varied little most of the time.

Smock Mill

Two smock mills with a stage in Greetsiel, Germany

The smock mill is a later development of the tower mill, where the tower is replaced by a wooden framework, called the "smock". The smock is commonly of octagonal plan, though examples with more, or fewer, sides exist. The smock is thatched, boarded or covered by other materials, such as slate, sheet metal, or tar paper. The lighter construction in comparison to tower mills make smock mills practical as drainage mills as these often had to be built in areas with unstable subsoil. Having originated as a drainage mill, smock mills are also used for a variety of purposes. When used in a built-up area it is often placed on a masonry base to raise it above the surrounding buildings.

Mechanics

Sails

Windmill in Kuremaa, Estonia

Common sails consist of a lattice framework on which a sailcloth is spread. The miller can adjust the amount of cloth spread according to the amount of wind available and power needed. In medieval mills, the sailcloth was wound in and out of a ladder type arrangement of sails. Postmedieval mill sails had a lattice framework over which the sailcloth was spread, while in colder climates, the cloth was replaced by wooden slats, which were easier to handle in freezing conditions. The jib sail is commonly found in Mediterranean countries, and consists of a simple triangle of cloth wound round a spar.

In all cases, the mill needs to be stopped to adjust the sails. Inventions in Great Britain in the late eighteenth and nineteenth centuries led to sails that automatically adjust to the wind speed without the need for the miller to intervene, culminating in patent sails invented by William Cubitt in 1807. In these sails, the cloth is replaced by a mechanism of connected shutters.

Windmills Zaanse Schans

In France, Pierre-Théophile Berton invented a system consisting of longitudinal wooden slats connected by a mechanism that lets the miller open them while the mill is turning. In the twentieth century, increased knowledge of aerodynamics from the development of the airplane led to further improvements in efficiency by German engineer Bilau and several Dutch millwrights. The majority of windmills have four sails. Multiple-sailed mills, with five, six or eight sails, were built in Great Britain (especially in and around the counties of Lincolnshire and Yorkshire), Germany, and less commonly elsewhere. Earlier multiple-sailed mills are found in Spain, Portugal,

Greece, parts of Romania, Bulgaria, and Russia. A mill with an even number of sails has the advantage of being able to run with a damaged sail and the one opposite removed without resulting in an unbalanced mill.

In the Netherlands the stationary position of the sails, i.e. when the mill is not working, has long been used to give signals. A slight tilt of the sails before the main building signals joy, while a tilt after the building signals mourning. Across the Netherlands, windmills were placed in mourning position in honor of the Dutch victims of the 2014 Malaysian Airlines Flight 17 shootdown.

Machinery

Gears inside a windmill convey power from the rotary motion of the sails to a mechanical device. The sails are carried on the horizontal windshaft. Windshafts can be wholly made of wood, or wood with a cast iron poll end (where the sails are mounted) or entirely of cast iron. The brake wheel is fitted onto the windshaft between the front and rear bearing. It has the brake around the outside of the rim and teeth in the side of the rim which drive the horizontal gearwheel called wallower on the top end of the vertical upright shaft. In grist mills, the great spur wheel, lower down the upright shaft, drives one or more stone nuts on the shafts driving each millstone. Post mills sometimes have a head and/or tail wheel driving the stone nuts directly, instead of the spur gear arrangement. Additional gear wheels drive a sack hoist or other machinery. The machinery differs if the windmill is used for other applications than milling grain. A drainage mill uses another set of gear wheels on the bottom end of the upright shaft to drive a scoop wheel or Archimedes' screw. Sawmills use a crankshaft to provide a reciprocating motion to the saws. Windmills have been used to power many other industrial processes, including papermills, threshing mills, and to process oil seeds, wool, paints and stone products.

Littlefield, Texas, claims the world's tallest windmill.

Interior view, Pantigo windmill, East Hampton, New York Historic American Buildings Survey

Windshaft, brake wheel, and brake blocks in smock mill d'Admiraal in Amsterdam

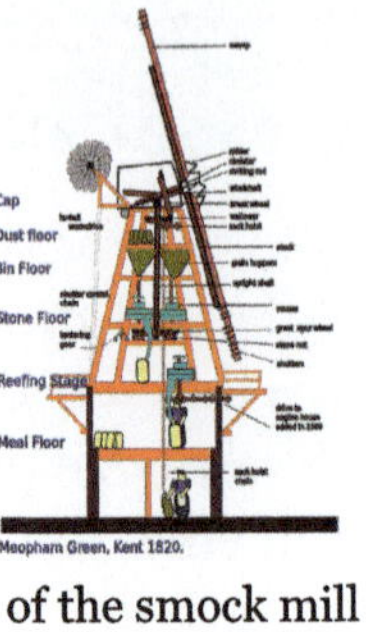

Diagram of the smock mill at Meopham, Kent

Spread and Decline

A windmill in Wales, United Kingdom. 1815.

Oilmill De Zoeker, paintmill De Kat and paltrok sawmill De Gekroonde Poelenburg at the Zaanse Schans

In the 14th century windmills became popular in Europe; the total number of wind-powered mills is estimated to have been around 200,000 at its peak in 1850, which is modest compared to some 500,000 waterwheels. Windmills were applied in regions where there was too little water, where rivers freeze in winter and in flat lands where the flow of the river was too slow to provide the required power. With the coming of the industrial revolution, the importance of wind and water as primary industrial energy sources declined and were eventually replaced by steam (in steam mills) and internal combustion engines, although windmills continued to be built in large numbers until late in the nineteenth century. More recently, windmills have been preserved for their historic value, in some cases as static exhibits when the antique machinery is too fragile to put in motion, and in other cases as fully working mills.

Of the 10,000 windmills in use in the Netherlands around 1850, about 1,000 are still standing. Most of these are being run by volunteers, though some grist mills are still operating commercially. Many of the drainage mills have been appointed as backup to the modern pumping stations. The Zaan district has been said to have been the first industrialized region of the world with around 600 operating wind-powered industries by the end of the eighteenth century. Economic fluctuations and the industrial revolution had a much greater impact on these industries than on grain and drainage mills, so only very few are left.

Don Quijote being struck by a windmill, illustration by Paul Gustave Louis Christophe Doré.

Construction of mills spread to the Cape Colony in the seventeenth century. The early tower mills did not survive the gales of the Cape Peninsula, so in 1717, the Heeren XVII sent carpenters, masons, and materials to construct a durable mill. The mill, completed in 1718, became known as the *Oude Molen* and was located between Pinelands Station and the Black River. Long since demolished, its name lives on as that of a Technical school in Pinelands. By 1863, Cape Town could boast 11 mills stretching from Paarden Eiland to Mowbray.

Wind Turbines

Rønland Offshore Wind Farm in Denmark

A wind turbine is a windmill-like structure specifically developed to generate electricity. They can be seen as the next step in the development of the windmill. The first wind turbines were built by the end of the nineteenth century by Prof James Blyth in Scotland (1887), Charles F. Brush in Cleveland, Ohio (1887–1888) and Poul la Cour in Denmark (1890s). La Cour's mill from 1896 later became the local powerplant of the village Askov. By 1908 there were 72 wind-driven electric generators in Denmark, ranging from 5 to 25 kW. By the 1930s, windmills were widely used to generate electricity on farms in the United States where distribution systems had not yet been installed, built by companies such as Jacobs Wind, Wincharger, Miller Airlite, Universal Aeroelectric, Paris-Dunn, Airline, and Winpower. The Dunlite Corporation produced turbines for similar locations in Australia.

A group of wind turbines in Zhangjiakou, China

Forerunners of modern horizontal-axis utility-scale wind generators were the WIME-3D in service in Balaklava USSR from 1931 until 1942, a 100-kW generator on a 30-m (100-ft) tower, the Smith-Putnam wind turbine built in 1941 on the mountain known as Grandpa's Knob in Castleton, Vermont, United States of 1.25 MW and the NASA wind turbines developed from 1974 through the mid-1980s. The development of these 13 experimental wind turbines pioneered many of the wind turbine design technologies in use today, including: steel tube towers, variable-speed generators, composite blade materials, and partial-span pitch control, as well as aerodynamic, structural, and acoustic engineering design capabilities. The modern wind power industry began in 1979 with the serial production of wind turbines by Danish manufacturers Kuriant, Vestas, Nordtank, and Bonus. These early turbines were small by today's standards, with capacities of 20–30 kW each. Since then, commercial turbines have increased greatly in size, with the Enercon E-126 capable of delivering up to 7 MW, while wind turbine production has expanded to many countries.

As the 21st century began, rising concerns over energy security, global warming, and eventual fossil fuel depletion led to an expansion of interest in all available forms of renewable energy. Worldwide, many thousands of wind turbines are now operating, with a total nameplate capacity of 194,400 MW. Europe accounted for 48% of the total in 2009.

Windpumps

Windpumps were used to pump water since at least the 9th century in what is now Afghanistan, Iran and Pakistan. The use of wind pumps became widespread across the Muslim world and later spread to China and India. Windmills were later used extensively in Europe, particularly in the Netherlands and the East Anglia area of Great Britain, from the late Middle Ages onwards, to drain land for agricultural or building purposes.

Aermotor-style windpump in South Dakota, USA

The American windmill, or wind engine, was invented by Daniel Halladay in 1854 and was used mostly for lifting water from wells. Larger versions were also used for tasks such as sawing wood, chopping hay, and shelling and grinding grain. In early California and some other states, the windmill was part of a self-contained domestic water system which included a hand-dug well and a wooden water tower supporting a redwood tank enclosed by wooden siding known as a tankhouse. During the late 19th century steel blades and steel towers replaced wooden construction. At their peak in 1930, an estimated 600,000 units were in use. Firms such as U.S. Wind Engine and Pump Company, Challenge Wind Mill and Feed Mill Company, Appleton Manufacturing Company, Star, Eclipse, Fairbanks-Morse, Dempster Mill Manufacturing Company and Aermotor became the main suppliers in North and South America. These windpumps are used extensively on farms and ranches in the United States, Canada, Southern Africa, and Australia. They feature a large number of blades, so they turn slowly with considerable torque in low winds and are self-regulating in high winds. A tower-top gearbox and crankshaft convert the rotary motion into reciprocating strokes carried downward through a rod to the pump cylinder below. Such mills pumped water and powered feed mills, saw mills, and agricultural machinery.

In Australia, the Griffiths Brothers at Toowoomba manufactured windmills of the American pattern from 1876, with the trade name Southern Cross Windmills in use from 1903. These became an icon of the Australian rural sector by utilizing the water of the Great Artesian Basin. Another well-known maker was Metters Ltd. of Adelaide, Perth and Sydney.

Wind Turbine

A wind turbine is a device that converts the wind's kinetic energy into electrical power.

Wind turbines are manufactured in a wide range of vertical and horizontal axis types. The smallest turbines are used for applications such as battery charging for auxiliary power for boats or caravans or to power traffic warning signs. Slightly larger turbines can be used for making contributions to a domestic power supply while selling unused power back to the utility supplier via the electrical grid. Arrays of large turbines, known as wind farms, are becoming an increasingly important source of intermittent renewable energy and are used by many countries as part of a strategy to reduce their reliance on fossil fuels.

Offshore wind farm, using 5 MW turbines REpower 5M in the North Sea off the coast of Belgium.

History

James Blyth's electricity-generating wind turbine, photographed in 1891

Wind turbines were used in Persia (present-day Iran) about 500–900 A.D. The windwheel of Hero of Alexandria marks one of the first known instances of wind powering a machine in history. However, the first known practical wind turbines were built in Sistan, an Eastern province of Iran, from the 7th century. These "Panemone" were vertical axle wind turbines, which had long vertical drive shafts with rectangular blades. Made of six to twelve sails covered in reed matting or cloth material, these wind turbines were used to grind grain or draw up water, and were used in the gristmilling and sugarcane industries.

Wind turbines first appeared in Europe during the Middle Ages. The first historical records of their use in England date to the 11th or 12th centuries and there are reports of German crusaders taking their windmill-making skills to Syria around 1190. By the 14th century, Dutch wind turbines were in use to drain areas of the Rhine delta. Advanced wind mills were described by Croatian inventor

Fausto Veranzio. In his book Machinae Novae (1595) he described vertical axis wind turbines with curved or V-shaped blades.

The first electricity-generating wind turbine was a battery charging machine installed in July 1887 by Scottish academic James Blyth to light his holiday home in Marykirk, Scotland. Some months later American inventor Charles F. Brush was able to build the first automatically operated wind turbine after consulting local University professors and colleagues Jacob S. Gibbs and Brinsley Coleberd and successfully getting the blueprints peer-reviewed for electricity production in Cleveland, Ohio. Although Blyth's turbine was considered uneconomical in the United Kingdom electricity generation by wind turbines was more cost effective in countries with widely scattered populations.

In Denmark by 1900, there were about 2500 windmills for mechanical loads such as pumps and mills, producing an estimated combined peak power of about 30 MW. The largest machines were on 24-meter (79 ft) towers with four-bladed 23-meter (75 ft) diameter rotors. By 1908 there were 72 wind-driven electric generators operating in the United States from 5 kW to 25 kW. Around the time of World War I, American windmill makers were producing 100,000 farm windmills each year, mostly for water-pumping.

By the 1930s, wind generators for electricity were common on farms, mostly in the United States where distribution systems had not yet been installed. In this period, high-tensile steel was cheap, and the generators were placed atop prefabricated open steel lattice towers.

A forerunner of modern horizontal-axis wind generators was in service at Yalta, USSR in 1931. This was a 100 kW generator on a 30-meter (98 ft) tower, connected to the local 6.3 kV distribution system. It was reported to have an annual capacity factor of 32 percent, not much different from current wind machines.

In the autumn of 1941, the first megawatt-class wind turbine was synchronized to a utility grid in Vermont. The Smith-Putnam wind turbine only ran for 1,100 hours before suffering a critical failure. The unit was not repaired, because of shortage of materials during the war.

The first utility grid-connected wind turbine to operate in the UK was built by John Brown & Company in 1951 in the Orkney Islands.

Despite these diverse developments, developments in fossil fuel systems almost entirely eliminated any wind turbine systems larger than supermicro size. In the early 1970s, however, anti-nuclear protests in Denmark spurred artisan mechanics to develop microturbines of 22 kW. Organizing owners into associations and co-operatives lead to the lobbying of the government and utilities and provided incentives for larger turbines throughout the 1980s and later. Local activists in Germany, nascent turbine manufacturers in Spain, and large investors in the United States in the early 1990s then lobbied for policies that stimulated the industry in those countries. Later companies formed in India and China. As of 2012, Danish company Vestas is the world's biggest wind-turbine manufacturer.

Resources

A quantitative measure of wind energy available at any location is called the Wind Power Density (WPD). It is a calculation of the mean annual power available per square meter of swept area of a turbine, and is tabulated for different heights above ground. Calculation of wind power density includes the effect of wind velocity and air density. Color-coded maps are prepared for a particular area described, for example, as "Mean Annual Power Density at 50 Metres". In the United States, the results of the above calculation are included in an index developed by the National Renewable Energy Laboratory and referred to as "NREL CLASS". The larger the WPD, the higher it is rated by class. Classes range from Class 1 (200 watts per square meter or less at 50 m altitude) to Class 7 (800 to 2000 watts per square m). Commercial wind farms generally are sited in Class 3 or higher areas, although isolated points in an otherwise Class 1 area may be practical to exploit.

Wind turbine in Germany

Wind turbines are classified by the wind speed they are designed for, from class I to class IV, with A or B referring to the turbulence.

Class	Avg Wind Speed (m/s)	Turbulence
IA	10	18%
IB	10	16%
IIA	8.5	18%
IIB	8.5	16%
IIIA	7.5	18%
IIIB	7.5	16%
IVA	6	18%
IVB	6	16%

Efficiency

Not all the energy of blowing wind can be used, but some small wind turbines are designed to work at low wind speeds.

Conservation of mass requires that the amount of air entering and exiting a turbine must be equal.

Accordingly, Betz's law gives the maximal achievable extraction of wind power by a wind turbine as 16/27 (59.3%) of the total kinetic energy of the air flowing through the turbine.

The maximum theoretical power output of a wind machine is thus 16/27 times the kinetic energy of the air passing through the effective disk area of the machine. If the effective area of the disk is A, and the wind velocity v, the maximum theoretical power output P is:

$$P = \frac{16}{27}\frac{1}{2}\rho v^3 A = \frac{8}{27}\rho v^3 A$$

where ρ is the air density.

As wind is free (no fuel cost), wind-to-rotor efficiency (including rotor blade friction and drag) is one of many aspects impacting the final price of wind power. Further inefficiencies, such as gearbox losses, generator and converter losses, reduce the power delivered by a wind turbine. To protect components from undue wear, extracted power is held constant above the rated operating speed as theoretical power increases at the cube of wind speed, further reducing theoretical efficiency. In 2001, commercial utility-connected turbines deliver 75% to 80% of the Betz limit of power extractable from the wind, at rated operating speed.

Efficiency can decrease slightly over time due to wear. Analysis of 3128 wind turbines older than 10 years in Denmark showed that half of the turbines had no decrease, while the other half saw a production decrease of 1.2% per year. Vertical turbine designs have much lower efficiency than standard horizontal designs.

Types

Wind turbines can rotate about either a horizontal or a vertical axis, the former being both older and more common. They can also include blades (transparent or not) or be bladeless. Vertical designs produce less power and are less common.

Horizontal Axis

Horizontal-axis wind turbines (HAWT) have the main rotor shaft and electrical generator at the top of a tower, and must be pointed into the wind. Small turbines are pointed by a simple wind vane, while large turbines generally use a wind sensor coupled with a servo motor. Most have a gearbox, which turns the slow rotation of the blades into a quicker rotation that is more suitable to drive an electrical generator.

Any solid object produces a wake behind it, leading to fatigue failures, so the turbine is usually positioned upwind of its supporting tower. Downwind machines have been built, because they don't need an additional mechanism for keeping them in line with the wind. In high winds, the blades can also be allowed to bend which reduces their swept area and thus their wind resistance. In upwind designs, turbine blades must be made stiff to prevent the blades from being pushed into the tower by high winds. Additionally, the blades are placed a considerable distance in front of the tower and are sometimes tilted forward into the wind a small amount.

Turbines used in wind farms for commercial production of electric power are usually three-bladed. These have low torque ripple, which contributes to good reliability. The blades are usually colored white for daytime visibility by aircraft and range in length from 20 to 80 meters (66 to 262 ft). The size and height of turbines increase year by year. Offshore wind turbines are built up to 8MW today and have a blade length up to 80m. Usual tubular steel towers of multi megawatt turbines have a height of 70 m to 120 m and in extremes up to 160 m.

The blades rotate at 10 to 22 revolutions per minute. At 22 rotations per minute the tip speed exceeds 90 meters per second (300 ft/s). Higher tip speeds means more noise and blade erosion. A gear box is commonly used for stepping up the speed of the generator, although designs may also use direct drive of an annular generator. Some models operate at constant speed, but more energy can be collected by variable-speed turbines which use a solid-state power converter to interface to the transmission system. All turbines are equipped with protective features to avoid damage at high wind speeds, by feathering the blades into the wind which ceases their rotation, supplemented by brakes.

Vertical Axis

Vertical-axis wind turbines (or VAWTs) have the main rotor shaft arranged vertically. One advantage of this arrangement is that the turbine does not need to be pointed into the wind to be effective, which is an advantage on a site where the wind direction is highly variable. It is also an advantage when the turbine is integrated into a building because it is inherently less steerable. Also, the generator and gearbox can be placed near the ground, using a direct drive from the rotor assembly to the ground-based gearbox, improving accessibility for maintenance. However, these designs produce much less energy averaged over time, which is a major drawback.

The key disadvantages include the relatively low rotational speed with the consequential higher torque and hence higher cost of the drive train, the inherently lower power coefficient, the 360-degree rotation of the aerofoil within the wind flow during each cycle and hence the highly dynamic loading on the blade, the pulsating torque generated by some rotor designs on the drive train, and the difficulty of modelling the wind flow accurately and hence the challenges of analysing and designing the rotor prior to fabricating a prototype.

When a turbine is mounted on a rooftop the building generally redirects wind over the roof and this can double the wind speed at the turbine. If the height of a rooftop mounted turbine tower is approximately 50% of the building height it is near the optimum for maximum wind energy and minimum wind turbulence. Wind speeds within the built environment are generally much lower than at exposed rural sites, noise may be a concern and an existing structure may not adequately resist the additional stress.

Offshore Horizontal Axis Wind Turbines (HAWTs) at Scroby Sands Wind Farm, UK

Subtypes of the vertical axis design include:

Darrieus Wind Turbine

"Eggbeater" turbines, or Darrieus turbines, were named after the French inventor, Georges Darrieus. They have good efficiency, but produce large torque ripple and cyclical stress on the tower, which contributes to poor reliability. They also generally require some external power source, or an additional Savonius rotor to start turning, because the starting torque is very low. The torque ripple is reduced by using three or more blades which results in greater solidity of the rotor. Solidity is measured by blade area divided by the rotor area. Newer Darrieus type turbines are not held up by guy-wires but have an external superstructure connected to the top bearing.

Giromill

A subtype of Darrieus turbine with straight, as opposed to curved, blades. The cycloturbine variety has variable pitch to reduce the torque pulsation and is self-starting. The advantages of variable pitch are: high starting torque; a wide, relatively flat torque curve; a higher coefficient of performance; more efficient operation in turbulent winds; and a lower blade speed ratio which lowers blade bending stresses. Straight, V, or curved blades may be used.

Savonius Wind Turbine

These are drag-type devices with two (or more) scoops that are used in anemometers, *Flettner* vents (commonly seen on bus and van roofs), and in some high-reliability low-efficiency power turbines. They are always self-starting if there are at least three scoops.

Twisted Savonius

Twisted Savonius is a modified savonius, with long helical scoops to provide smooth torque. This is often used as a rooftop windturbine and has even been adapted for ships.

Another type of vertical axis is the Parallel turbine, which is similar to the crossflow fan or centrifugal fan. It uses the ground effect. Vertical axis turbines of this type have been tried for many years: a unit producing 10 kW was built by Israeli wind pioneer Bruce Brill in the 1980s.

Design and Construction

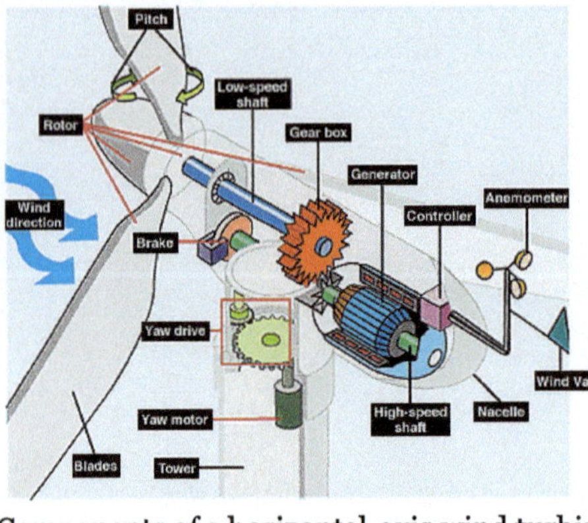

Components of a horizontal-axis wind turbine

Inside view of a wind turbine tower, showing the tendon cables.

Wind turbines are designed, using a range of computer modelling techniques, to exploit the wind energy that exists at a location. For example, Aerodynamic modeling is used to determine the optimum tower height, control systems, number of blades and blade shape.

Wind turbines convert wind energy to electricity for distribution. Conventional horizontal axis turbines can be divided into three components:

- The rotor component, which is approximately 20% of the wind turbine cost, includes the blades for converting wind energy to low speed rotational energy.
- The generator component, which is approximately 34% of the wind turbine cost, includes the electrical generator, the control electronics, and most likely a gearbox (e.g. planetary gearbox), adjustable-speed drive or continuously variable transmission component for converting the low speed incoming rotation to high speed rotation suitable for generating electricity.
- The structural support component, which is approximately 15% of the wind turbine cost, includes the tower and rotor yaw mechanism.

A 1.5 MW wind turbine of a type frequently seen in the United States has a tower 80 meters (260 ft) high. The rotor assembly (blades and hub) weighs 22,000 kilograms (48,000 lb). The nacelle, which contains the generator component, weighs 52,000 kilograms (115,000 lb). The concrete base for the tower is constructed using 26,000 kilograms (58,000 lb) of reinforcing steel and contains 190 cubic meters (250 cu yd) of concrete. The base is 15 meters (50 ft) in diameter and 2.4 meters (8 ft) thick near the center.

Among all renewable energy systems wind turbines have the highest effective intensity of power-harvesting surface because turbine blades not only harvest wind power, but also concentrate it.

Unconventional Designs

An E-66 wind turbine in the Windpark Holtriem, Germany, has an observation deck for visitors. Another turbine of the same type with an observation deck is located in Swaffham, England. Airborne wind turbine designs have been proposed and developed for many years but have yet to produce significant amounts of energy. In principle, wind turbines may also be used in conjunction with a large vertical solar updraft tower to extract the energy due to air heated by the sun.

Counter rotating wind turbine (dual rotor)

The corkscrew shaped wind turbine at Progressive Field in Cleveland, Ohio

Wind turbines which utilise the Magnus effect have been developed.

A ram air turbine (RAT) is a special kind of small turbine that is fitted to some aircraft. When deployed, the RAT is spun by the airstream going past the aircraft and can provide power for the most essential systems if there is a loss of all on-board electrical power, as in the case of the "Gimli Glider".

The two-bladed SCD 6MW offshore turbine designed by aerodyn Energiesysteme and built by MingYang Wind Power has a helideck for helicopters on top of its nacelle. The prototype was erected in 2014 in Rudong, China.

Turbine Monitoring and Diagnostics

Due to data transmission problems, structural health monitoring of wind turbines is usually performed using several accelerometers and strain gages attached to the nacelle to monitor the gearbox and equipments. Currently, digital image correlation and stereophotogrammetry are used to measure dynamics of wind turbine blades. These methods usually measure displacement and strain to identify location of defects. Dynamic characteristics of non-rotating wind turbines have been measured using digital image correlation and photogrammetry. Three dimensional point tracking has also been used to measure rotating dynamics of wind turbines.

Materials and Durability

Materials that are typically used for the rotor blades in wind turbines are composites, as they tend to have a high stiffness, high strength, high fatigue resistance, and low weight. Typical resins used for these composites include polyester and epoxy, while glass and carbon fibers have been used for the reinforcing material. Construction may use manual layup techniques or composite resin injection molding. As the price of glass fibers is only about one tenth the price of carbon fiber, glass fiber is still dominant.

As competition in the wind market increases, companies are seeking ways to draw greater efficiency from their designs. One of the predominant ways wind turbines have gained performance is by increasing rotor diameters, and thus blade length. Retrofitting current turbines with larger blades mitigates the need and risks associated with a system-level redesign. By incorporating carbon fiber into parts of existing blade systems, manufacturers may increase the length of the blades without increasing their overall weight. For instance, the spar cap, a structural element of a turbine blade, commonly experiences high tensile loading, making it an ideal candidate to utilize the enhanced tensile properties of carbon fiber in comparison to glass fiber. Higher stiffness and lower density translates to thinner, lighter blades offering equivalent performance. In a 10-MW turbine—which will become more common in offshore systems by 2021—blade lengths may reach over 100 m and weigh up to 50 metric tons when fabricated out of glass fiber. A switch to carbon fiber in the structural spar of the blade yields weight savings of 20 to 30 percent, or approximately 15 metric tons. The compressive properties of carbon fiber do not differ significantly from those of glass fiber. It is therefore economical to replace glass fiber components under compression with carbon fiber components.

While the material cost is significantly higher for all-glass fiber blades than for hybrid glass/carbon fiber blades, there is a potential for tremendous savings in manufacturing costs when labor price is considered. Utilizing carbon fiber enables for simpler designs that use less raw material. The chief manufacturing process in blade fabrication is the layering of plies. By reducing the number of layers of plies, as is enabled by thinner blade design, the cost of labor may be decreased, and in some cases, equate to the cost of labor for glass fiber blades.

Materials for wind turbine parts other than the rotor blades (including the rotor hub, gearbox, frame, and tower) are largely composed of steel. Modern turbines uses a couple of tonnes of copper for generators, cables and such. Smaller wind turbines have begun incorporating more aluminum based alloys into these components in an effort to make the turbines more lightweight and efficient, and may continue to be used increasingly if fatigue and strength properties can be improved. Prestressed concrete has been increasingly used for the material of the tower, but still, requires much reinforcing steel to meet the strength requirement of the turbine. Additionally, step-up gearboxes are being increasingly replaced with variable speed generators, increasing the demand for magnetic materials in wind turbines., In particular, this would require an increased supply of the rare earth metal neodymium. Reliance on rare earth minerals for components has risked expense and price volatility as China has been main producer of rare earth minerals (96% in

2009) and had been reducing its export quotas of these materials. In recent years, however, other producers have increased production of rare earth minerals and China has removed its reduced export quota on rare earths leading to an increased supply and decreased cost of rare earth minerals, increasing the viability of the implementation of variable speed generators in wind turbines on a large scale.

Wind Turbines on Public Display

A few localities have exploited the attention-getting nature of wind turbines by placing them on public display, either with visitor centers around their bases, or with viewing areas farther away. The wind turbines are generally of conventional horizontal-axis, three-bladed design, and generate power to feed electrical grids, but they also serve the unconventional roles of technology demonstration, public relations, and education.

The Nordex N50 wind turbine and visitor centre of Lamma Winds in Hong Kong, China

Small Wind Turbines

Small wind turbines may be used for a variety of applications including on- or off-grid residences, telecom towers, offshore platforms, rural schools and clinics, remote monitoring and other purposes that require energy where there is no electric grid, or where the grid is unstable. Small wind turbines may be as small as a fifty-watt generator for boat or caravan use. Hybrid solar and wind powered units are increasingly being used for traffic signage, particularly in rural locations, as they avoid the need to lay long cables from the nearest mains connection point. The U.S. Department of Energy's National Renewable Energy Laboratory (NREL) defines small wind turbines as those smaller than or equal to 100 kilowatts. Small units often have direct drive generators, direct current output, aeroelastic blades, lifetime bearings and use a vane to point into the wind.

Larger, more costly turbines generally have geared power trains, alternating current output, flaps and are actively pointed into the wind. Direct drive generators and aeroelastic blades for large wind turbines are being researched.

A small Quietrevolution QR5 Gorlov type vertical axis wind turbine in Bristol, England. Measuring 3 m in diameter and 5 m high, it has a nameplate rating of 6.5 kW to the grid.

Wind Turbine Spacing

On most horizontal windturbine farms, a spacing of about 6–10 times the rotor diameter is often upheld. However, for large wind farms distances of about 15 rotor diameters should be more economically optimal, taking into account typical wind turbine and land costs. This conclusion has been reached by research conducted by Charles Meneveau of the Johns Hopkins University, and Johan Meyers of Leuven University in Belgium, based on computer simulations that take into account the detailed interactions among wind turbines (wakes) as well as with the entire turbulent atmospheric boundary layer. Moreover, recent research by John Dabiri of Caltech suggests that vertical wind turbines may be placed much more closely together so long as an alternating pattern of rotation is created allowing blades of neighbouring turbines to move in the same direction as they approach one another.

Operability

Maintenance

Wind turbines need regular maintenance to stay reliable and available, reaching 98%.

Modern turbines usually have a small onboard crane for hoisting maintenance tools and minor components. However, large heavy components like generator, gearbox, blades and so on are rarely replaced and a heavy lift external crane is needed in those cases. If the turbine has a difficult access road, a containerized crane can be lifted up by the internal crane to provide heavier lifting.

Repowering

Installation of new wind turbines can be controversial. An alternative is repowering, where existing wind turbines are replaced with bigger, more powerful ones, sometimes in smaller numbers while keeping or increasing capacity.

Demolition

Older turbines were in some early cases not required to be removed when reaching the end of their life. Some still stand, waiting to be recycled or repowered.

A demolition industry develops to recycle offshore turbines at a cost of DKK 2–4 million per MW, to be guaranteed by the owner.

Advantages and Disadvantages

Advantages

Wind turbines are generally inexpensive. They will cost between two and six cents per kilowatt hour, which is one of the lowest-priced renewable energy sources in today's world. And as technology needed for wind turbines continues to improve, the prices will decrease as well. In addition, there is no competitive market for wind energy, as it does not cost money to get ahold of wind. The main cost of wind turbines are the installation process. The average cost is between $48,000 and $65,000 to install. However, the energy harvested from the turbine will offset the installation cost, as well as provide virtually free energy for years after.

Wind turbines provide a clean energy source, emitting no greenhouse gases and no waste product. Over 1,500 tons of carbon dioxide per year can be eliminated by using a one megawatt turbine instead of one megawatt of energy from a fossil fuel. Being environmentally friendly and green is a large advantage of wind turbines.

Wind turbines are also quite efficient. Wind farms can generate between 17 and 39 times as much power as they consume, and in the United States alone, wind turbines have produced about 16 billion kilowatt-hours of energy per year.

Disadvantages

Wind turbines can be very large, reaching over 400 feet tall and with blades 50 yards long. This can make them visual pollution, and people have often complained of this. Their size can also make them harmful to surrounding wildlife. Thousands of birds, including rare species, have been killed by the blades of wind turbines or by being electrocuted by the power lines.

Energy harnessed by wind turbines can be unreliable at times. Wind is often not available when needed, as wind can fluctuate. Turbines can be placed on ridges or bluffs to maximize the access of wind they have, but this also limits the locations where they can be placed. In this way, wind energy is not a particularly reliable source of energy.

Records

Fuhrländer Wind Turbine Laasow, in Brandenburg, Germany, among the world's tallest wind turbines

Éole, the largest vertical axis wind turbine, in Cap-Chat, Quebec, Canada

Largest Capacity Conventional Drive

The Vestas V164 has a rated capacity of 8 MW, has an overall height of 220 m (722 ft), a diameter of 164 m (538 ft), is for offshore use, and is the world's largest-capacity wind turbine since its introduction in 2014. The conventional drive train consist of a main gear-box and a medium speed PM generator. Prototype installed in 2014 at the National Test Center Denmark nearby Østerild. Series production starts end of 2015.

Largest Capacity Direct Drive

The Enercon E-126 with 7.58 MW and 127 m rotor diameter is the largest direct drive turbine. It's only for onshore use. The turbine has parted rotor blades with 2 sections for transport. In July 2016, Siemens upgraded its 7 to 8 MW.

Largest Vertical-axis

Le Nordais wind farm in Cap-Chat, Quebec has a vertical axis wind turbine (VAWT) named Éole, which is the world's largest at 110 m. It has a nameplate capacity of 3.8 MW.

Largest 1-bladed Turbine

Riva Calzoni M33 was a single-bladed wind turbine with 350 kW, designed and built In Bologna in 1993.

Largest 2-bladed Turbine

The biggest 2-bladed turbine is built by Mingyang Wind Power in 2013. It is a SCD6.5MW offshore downwind turbine, designed by aerodyn Energiesysteme.

Largest Swept Area

The turbine with the largest swept area is the Samsung S7.0–171, with a diameter of 171 m, giving a total sweep of 22966 m^2.

Tallest

A Nordex 3.3 MW was installed in July 2016. It has a total height of 230m, and a hub height of 164m on 100m concrete tower bottom with steel tubes on top (hybrid tower).

Vestas V164 was the tallest wind turbine, standing in Østerild, Denmark, 220 meters tall, constructed in 2014. It has a steel tube tower.

Highest Tower

Fuhrländer installed a 2.5MW turbine on a 160m lattice tower in 2003 (Fuhrländer Wind Turbine Laasow and Nowy Tomyśl Wind Turbines).

Most Rotors

Lagerwey has build Four-in-One, a multi rotor wind turbine with one tower and four rotors near Maasvlakte. In April 2016, Vestas installed a 900 kW quadrotor test wind turbine at Risø, made from 4 recycled 225 kW V29 turbines.

Most Productive

Four turbines at Rønland Offshore Wind Farm in Denmark share the record for the most productive wind turbines, with each having generated 63.2 GWh by June 2010.

Highest-situated

Since 2013 the world's highest-situated wind turbine was made and installed by WindAid and is located at the base of the Pastoruri Glacier in Peru at 4,877 meters (16,001 ft) above sea level. The site uses the WindAid 2.5 kW wind generator to supply power to a small rural community of micro entrepreneurs who cater to the tourists who come to the Pastoruri glacier.

Largest floating Wind Turbine

The world's largest—and also the first operational deep-water *large-capacity*—floating wind turbine is the 2.3 MW Hywind currently operating 10 kilometers (6.2 mi) offshore in 220-meter-deep water, southwest of Karmøy, Norway. The turbine began operating in September 2009 and utilizes a Siemens 2.3 MW turbine.

History of Wind-Mills

The wind is a by-product of solar energy. Approximately 2% of the sun's energy reaching the earth is converted into wind energy. The surface of the earth heats and cools unevenly, creating atmospheric pressure zones that make air flow from high- to low-pressure areas.

The wind has played an important role in the history of human civilization. The first known use of wind dates back 5,000 years to Egypt, where boats used sails to travel from shore to shore. The first true windmill, a machine with vanes attached to an axis to produce circular motion, may have been built as early as 2000 B.C. in ancient Babylon. By the 10th century A.D., windmills with wind-catching surfaces having 16 feet length and 30 feet height were grinding grain in the areas in eastern Iran and Afghanistan.

The earliest written references to working wind machines in western world date from the 12th century. These too were used for milling grain. It was not until a few hundred years later that windmills were modified to pump water and reclaim much of Holland from the sea.

The multi-vane "farm windmill" of the American Midwest and West was invented in the United States during the latter half of the l9th century. In 1889 there were 77 windmill factories in the United States, and by the turn of the century, windmills had become a major American export. Until the diesel engine came along, many transcontinental rail routes in the U.S. depended on large multi-vane windmills to pump water for steam locomotives.

Farm windmills are still being produced and used, though in reduced numbers. They are best suited for pumping ground water in small quantities to livestock water tanks. In the 1930s and 1940s, hundreds of thousands of electricity producing wind turbines were built in the U.S. They had two or three thin blades which rotated at high speeds to drive electrical generators. These wind turbines provided electricity to farms beyond the reach of power lines and were typically used to charge storage batteries, operate radio receivers and power a light bulb. By the early 1950s, however, the extension of the central power grid to nearly every American household, via the Rural Electrification Administration, eliminated the market for these machines. Wind turbine development lay nearly dormant for the next 20 years.

A typical modern windmill looks as shown in the following figure. The wind-mill contains three blades about a horizontal axis installed on a tower. A turbine connected to a generator is fixed about the horizontal axis.

Like the weather in general, the wind can be unpredictable. It varies from place to place, and from moment to moment. Because it is invisible, it is not easily measured without special instruments. Wind velocity is affected by the trees, buildings, hills and valleys around us. Wind is a diffuse energy source that cannot be contained or stored for use elsewhere or at another time.

Classification of Wind-mills

Wind turbines are classified into two general types: Horizontal axis and Vertical axis. A horizontal axis machine has its blades rotating on an axis parallel to the ground as shown in the above figure. A vertical axis machine has its blades rotating on an axis perpendicular to the ground. There are a number of available designs for both and each type has certain advantages and disadvantages. However, compared with the horizontal axis type, very few vertical axis machines are available commercially.

Horizontal Axis

This is the most common wind turbine design. In addition to being parallel to the ground, the axis of blade rotation is parallel to the wind flow. Some machines are designed to operate in an upwind mode, with the blades upwind of the tower. In this case, a tail vane is usually used to keep the blades facing into the wind. Other designs operate in a downwind mode so that the wind passes the tower before striking the blades. Without a tail vane, the machine rotor naturally tracks the wind in a downwind mode. Some very large wind turbines use a motor-driven mechanism that turns the machine in response to a wind direction sensor mounted on the tower. Commonly found horizontal axis wind mills are aero-turbine mill with 35% efficiency and farm mills with 15% efficiency.

Vertical Axis

Although vertical axis wind turbines have existed for centuries, they are not as common as their horizontal counterparts. The main reason for this is that they do not take advantage of the higher wind speeds at higher elevations above the ground as well as horizontal axis turbines. The basic vertical axis designs are the Darrieus, which has curved blades and efficiency of 35%, the Giromill, which has straight blades, and efficiency of 35%, and the Savonius, which uses scoops to catch the wind and the efficiency of 30%. A vertical axis machine need not be oriented with respect to wind direction. Because the shaft is vertical, the transmission and generator can be mounted at ground level allowing easier servicing and a lighter weight, lower cost tower. Although vertical axis wind turbines have these advantages, their designs are not as efficient at collecting energy from the wind as are the horizontal machine designs. The following figures show all the above mentioned mills.

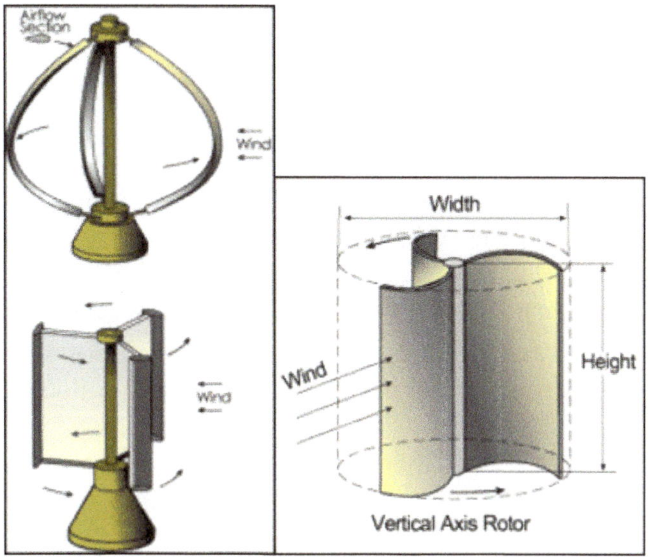

There is one more type of wind-mill called Cyclo-gyro wind-mill with very high efficiency of about 60%. However, it is not very stable and is very sensitive to wind direction. It is also very complex to build.

Main Components of a wind-mill

Following figure shows typical components of a horizontal axis wind mill.

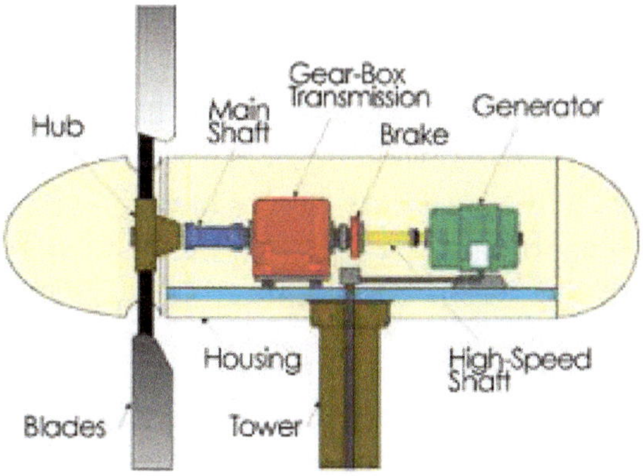

Rotor

The portion of the wind turbine that collects energy from the wind is called the rotor. The rotor usually consists of two or more wooden, fiberglass or metal blades which rotate about an axis (horizontal or vertical) at a rate determined by the wind speed and the shape of the blades. The blades are attached to the hub, which in turn is attached to the main shaft.

Drag Design

Blade designs operate on either the principle of drag or lift. For the drag design, the wind literally pushes the blades out of the way. Drag powered wind turbines are characterized by slower rotational speeds and high torque capabilities. They are useful for the pumping, sawing or grinding work. For example, a farm-type windmill must develop high torque at start-up in order to pump, or lift, water from a deep well.

Lift Design

The lift blade design employs the same principle that enables airplanes, kites and birds to fly. The blade is essentially an airfoil, or wing. When air flows past the blade, a wind speed and pressure differential is created between the upper and lower blade surfaces. The pressure at the lower surface is greater and thus acts to "lift" the blade. When blades are attached to a central axis, like a wind turbine rotor, the lift is translated into rotational motion. Lift-powered wind turbines have much higher rotational speeds than drag types and therefore well suited for electricity generation.

Following figure gives an idea about the drag and lift principle.

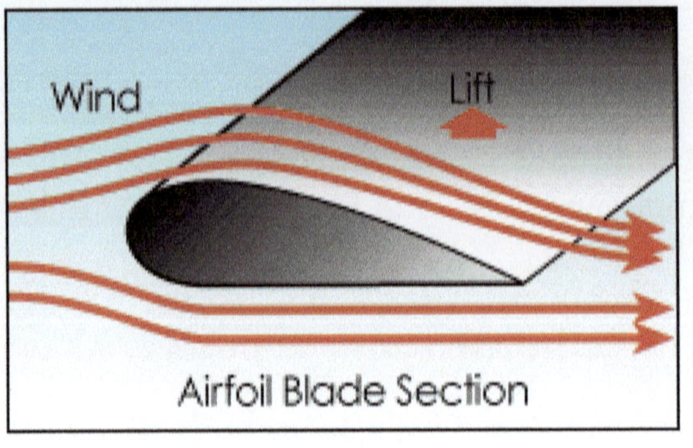

Tip Speed Ratio

The tip-speed is the ratio of the rotational speed of the blade to the wind speed. The larger this ratio, the faster the rotation of the wind turbine rotor at a given wind speed. Electricity generation requires high rotational speeds. Lift-type wind turbines have maximum tip-speed ratios of around 10, while drag-type ratios are approximately 1. Given the high rotational speed requirements of electrical generators, it is clear that the lift-type wind turbine is most practical for this application.

The number of blades that make up a rotor and the total area they cover affect wind turbine performance. For a lift-type rotor to function effectively, the wind must flow smoothly over the blades. To avoid turbulence, spacing between blades should be great enough so that one blade will not encounter the disturbed, weaker air flow caused by the blade which passed before it. It is because of this requirement that most wind turbines have only two or three blades on their rotors.

Generator

The generator is what converts the turning motion of a wind turbine's blades into electricity. Inside this component, coils of wire are rotated in a magnetic field to produce electricity. Different generator designs produce either alternating current (AC) or direct current (DC), and they are available in a large range of output power ratings. The generator's rating, or size, is dependent on the length of the wind turbine's blades because more energy is captured by longer blades.

It is important to select the right type of generator to match intended use. Most home and office appliances operate on 240 volt, 50 cycles AC. Some appliances can operate on either AC or DC, such as light bulbs and resistance heaters, and many others can be adapted to run on DC. Storage systems using batteries store DC and usually are configured at voltages of between 12 volts and 120 volts.

Generators that produce AC are generally equipped with features to produce the correct voltage of 240 V and constant frequency 50 cycles of electricity, even when the wind speed is fluctuating.

DC generators are normally used in battery charging applications and for operating DC appliances and machinery. They also can be used to produce AC electricity with the use of an inverter, which converts DC to AC.

Transmission

The number of revolutions per minute (rpm) of a wind turbine rotor can range between 40 rpm and 400 rpm, depending on the model and the wind speed. Generators typically require rpm's of 1,200 to 1,800. As a result, most wind turbines require a gear-box transmission to increase the rotation of the generator to the speeds necessary for efficient electricity production. Some DC-type wind turbines do not use transmissions. Instead, they have a direct link between the rotor and generator. These are known as direct drive systems. Without a transmission, wind turbine complexity and maintenance requirements are reduced, but a much larger generator is required to deliver the same power output as the AC-type wind turbines.

Tower

The tower on which a wind turbine is mounted is not just a support structure. It also raises the wind turbine so that its blades safely clear the ground and so it can reach the stronger winds at higher elevations. Maximum tower height is optional in most cases, except where zoning restrictions apply. The decision of what height tower to use will be based on the cost of taller towers versus the value of the increase in energy production resulting from their use. Studies have shown that the added cost of increasing tower height is often justified by the added power generated from the stronger winds. Larger wind turbines are usually mounted on towers ranging from 40 to 70 meters tall.

Towers for small wind systems are generally "guyed" designs. This means that there are guy wires anchored to the ground on three or four sides of the tower to hold it erect. These towers cost less than freestanding towers, but require more land area to anchor the guy wires. Some of these guyed towers are erected by tilting them up. This operation can be quickly accomplished using only a winch, with the turbine already mounted to the tower top. This simplifies not only installation, but maintenance as well. Towers can be constructed of a simple tube, a wooden pole or a lattice of tubes, rods, and angle iron. Large wind turbines may be mounted on lattice towers, tube towers or guyed tilt-up towers.

Towers must be strong enough to support the wind turbine and to sustain vibration, wind loading and the overall weather elements for the lifetime of the wind turbine. Their costs will vary widely as a function of design and height.

Operating Characteristics of Wind Mills

All wind machines share certain operating characteristics, such as cut-in, rated and cut- out wind speeds.

Cut-in Speed

Cut-in speed is the minimum wind speed at which the blades will turn and generate usable power. This wind speed is typically between 10 and 16 kmph.

Rated Speed

The rated speed is the minimum wind speed at which the wind turbine will generate its designated rated power. For example, a "10 kilowatt" wind turbine may not generate 10 kilowatts until wind speeds reach 40 kmph. Rated speed for most machines is in the range of 40 to 55 kmph. At wind speeds between cut-in and rated, the power output from a wind turbine increases as the wind increases. The output of most machines levels off above the rated speed. Most manufacturers provide graphs, called "power curves," showing how their wind turbine output varies with wind speed.

Cut-out Speed

At very high wind speeds, typically between 72 and 128 kmph, most wind turbines cease power generation and shut down. The wind speed at which shut down occurs is called the cut-out speed. Having a cut-out speed is a safety feature which protects the wind turbine from damage. Shut down may occur in one of several ways. In some machines an automatic brake is activated by a wind speed sensor. Some machines twist or "pitch" the blades to spill the wind. Still others use "spoilers," drag flaps mounted on the blades or the hub which are automatically activated by high rotor rpm's, or mechanically activated by a spring loaded device which turns the machine sideways to the wind stream. Normal wind turbine operation usually resumes when the wind drops back to a safe level.

Betz Limit

It is the flow of air over the blades and through the rotor area that makes a wind turbine function. The wind turbine extracts energy by slowing the wind down. The theoretical maximum amount of energy in the wind that can be collected by a wind turbine's rotor is approximately 59%. This value is known as the Betz limit. If the blades were 100% efficient, a wind turbine would not work because the air, having given up all its energy, would entirely stop. In practice, the collection efficiency of a rotor is not as high as 59%. A more typical efficiency is 35% to 45%. A complete wind energy system, including rotor, transmission, generator, storage and other devices, which all have less than perfect efficiencies, will deliver between 10% and 30% of the original energy available in the wind.

Wind-turbine Aerodynamics

The primary application of wind turbines is to generate energy using the wind. Hence, the aerodynamics is a very important aspect of wind turbines. Like most machines, there are many different types of wind turbines, all of them based on different energy extraction concepts.

Though the details of the aerodynamics depend very much on the topology, some fundamental concepts apply to all turbines. Every topology has a maximum power for a given flow, and some topologies are better than others. The method used to extract power has a strong influence on this. In general, all turbines may be grouped as being either lift-based, or drag-based; the former being more efficient. The difference between these groups is the aerodynamic force that is used to extract the energy.

Wind-turbine blades awaiting installation in laydown yard.

The most common topology is the horizontal-axis wind turbine (HAWT). It is a lift-based wind turbine with very good performance. Accordingly, it is a popular choice for commercial applications and much research has been applied to this turbine. Despite being a popular lift-based alternative in the latter part of the 20th century, the Darrieus wind turbine is rarely used today. The Savonius wind turbine is the most common drag type turbine. Despite its low efficiency, it remains in use because of its robustness and simplicity to build and maintain.

General Aerodynamic Considerations

The governing equation for power extraction is stated below:

$$P = \vec{F} \cdot \vec{v} \qquad (1)$$

Where: P is the power, F is the force vector, and v is the velocity of the moving wind turbine part.

The force F is generated by the wind's interaction with the blade. The magnitude and distribution of this force is the primary focus of wind-turbine aerodynamics. The most familiar type of aerodynamic force is drag. The direction of the drag force is parallel to the relative wind. Typically, the wind turbine parts are moving, altering the flow around the part. An example of relative wind is the wind one would feel cycling on a calm day.

To extract power, the turbine part must move in the direction of the net force. In the drag force case, the relative wind speed decreases subsequently, and so does the drag force. The relative wind aspect dramatically limits the maximum power that can be extracted by a drag-based wind turbine. Lift-based wind turbines typically have lifting surfaces moving perpendicular to the flow. Here, the relative wind does not decrease; rather, it increases with rotor speed. Thus, the maximum power limits of these machines are much higher than those of drag-based machines.

Characteristic Parameters

Wind turbines come in a variety of sizes. Once in operation, a wind turbine experiences a wide range of conditions. This variability complicates the comparison of different types of turbines. To deal with this, nondimensionalization is applied to various qualities. Nondimensionalization allows one to make comparisons between different turbines, without having to consider the effect of things like size and wind conditions from the comparison. One of the qualities of nondimensionalization is that though geometrically similar turbines will produce the same non-dimensional results, other factors (difference in scale, wind properties) cause them to produce very different dimensional properties.

The coefficient of power is the most important variable in wind-turbine aerodynamics. The Buckingham π theorem can be applied to show that the non-dimensional variable for power is given by the equation below. This equation is similar to efficiency, so values between 0 and less than 1 are typical. However, this is not exactly the same as efficiency and thus in practice, some turbines can exhibit greater than unity power coefficients. In these circumstances, one cannot conclude the first law of thermodynamics is violated because this is not an efficiency term by the strict definition of efficiency.

$$C_P = \frac{P}{\frac{1}{2}\rho A V^3} \qquad (CP)$$

Where: C_P is the coefficient of power, ρ is the air density, A is the area of the wind turbine, and V is the wind speed.

The thrust coefficient is another important dimensionless number in wind turbine aerodynamics.

$$C_T = \frac{T}{\frac{1}{2}\rho A V^2} \qquad (CT)$$

Equation shows two important dependents. The first is the speed (U) that the machine is going at. The speed at the tip of the blade is usually used for this purpose, and is written as the product of the blade radius and the rotational speed of the wind ("U=\omega r", where "\omega" is the rotational velocity in radians/second). This variable is nondimensionalized by the wind speed, to obtain the speed ratio:

$$\lambda = \frac{U}{V} \qquad (SpeedRatio)$$

The force vector is not straightforward, as stated earlier there are two types of aerodynamic forces, lift and drag. Accordingly, there are two non-dimensional parameters. However both variables are non-dimensionalized in a similar way. The formula for lift is given below, the formula for drag is given after:

$$C_L = \frac{L}{\frac{1}{2}\rho A W^2} \qquad (\boldsymbol{CL})$$

$$C_D = \frac{D}{\frac{1}{2}\rho A W^2} \qquad (\boldsymbol{CD})$$

Where: C_L is the lift coefficient, C_D is the drag coefficient, W is the relative wind as experienced by the wind turbine blade, and A is the area. Note that A may not be the same area used in the power non-dimensionalization of power.

The aerodynamic forces have a dependency on W, this speed is the relative speed and it is given by the equation below. Note that this is vector subtraction.

$$\vec{W} = \vec{V} - \vec{U} \qquad (\boldsymbol{RelativeSpeed})$$

Drag- versus lift-based Machines

All wind turbines extract energy from the wind through aerodynamic forces. There are two important aerodynamic forces: drag and lift. Drag applies a force on the body in the direction of the relative flow, while lift applies a force perpendicular to the relative flow. Many machine topologies could be classified by the primary force used to extract the energy. For example, a Savonious wind turbine is a drag-based machine, while a Darrieus wind turbine and conventional horizontal axis wind turbines are lift-based machines. Drag-based machines are conceptually simple, yet suffer from poor efficiency. Efficiency in this analysis is based on the power extracted vs. the plan-form area. Considering that the wind is free, but the blade materials are not, a plan-form-based definition of efficiency is more appropriate.

The analysis is focused on comparing the maximum power extraction modes and nothing else. Accordingly, several idealizations are made to simplify the analysis, further considerations are required to apply this analysis to real turbines. For example, in this comparison the effects of axial momentum theory are ignored. Axial momentum theory demonstrates how the wind turbine imparts an influence on the wind which in-turn decelerates the flow and limits the maximum power. For more details see Betz's law. Since this effect is the same for both lift and drag-based machines it can be ignored for comparison purposes. The topology of the machine can introduce additional losses, for example trailing vorticity in horizontal axis machines degrade the performance at the tip. Typically these losses are minor and can be ignored in this analysis (for example tip loss effects can be reduced with using high aspect-ratio blades).

Maximum Power of a drag-based Wind Turbine

Equation (1) will be the starting point in this derivation. Equation (CD) is used to define the force, and equation (RelativeSpeed) is used for the relative speed. These substitutions give the following formula for power.

$$P = \frac{1}{2}\rho A C_D \left(UV^2 - 2VU^2 + U^3\right) \qquad \textbf{\textit{(DragPower)}}$$

The formulas (CP) and (SpeedRatio) are applied to express (DragPower) in nondimensional form:

$$C_P = C_D \left(\lambda - 2\lambda^2 + \lambda^3\right) \qquad \textbf{\textit{(DragCP)}}$$

It can be shown through calculus that equation (DragCP) achieves a maximum at $\lambda = 1/3$. By inspection one can see that equation (DragPower) will achieve larger values for $\lambda > 1$. In these circumstances, the scalar product in equation (1) makes the result negative. Thus, one can conclude that the maximum power is given by:

$$C_P = \frac{4}{27} C_D$$

Experimentally it has been determined that a large C_D is 1.2, thus the maximum C_P is approximately 0.1778.

Maximum Power of a lift-based Wind Turbine

The derivation for the maximum power of a lift-based machine is similar, with some modifications. First we must recognize that drag is always present, and thus cannot be ignored. It will be shown that neglecting drag leads to a final solution of infinite power. This result is clearly invalid, hence we will proceed with drag. As before, equations (1), (CD) and (RelativeSpeed) will be used along with (CL) to define the power below expression.

$$P = \frac{1}{2}\rho A \sqrt{U^2 + V^2} \left(C_L UV - C_D U^2\right) \qquad \textbf{\textit{(LiftPower)}}$$

Similarly, this is non-dimensionalized with equations (CP) and (SpeedRatio). However, in this derivation the parameter $\gamma = C_D / C_L$ is also used:

$$C_P = C_L \sqrt{1 + \lambda^2} \left(\lambda - \gamma\lambda^2\right) \qquad \textbf{\textit{(LiftCP)}}$$

Solving the optimal speed ratio is complicated by the dependency on γ and the fact that the optimal speed ratio is a solution to a cubic polynomial. Numerical methods can then be applied to determine this solution and the corresponding C_P solution for a range of γ results. Some sample solutions are given in the table below.

γ	**Optimal** λ	**Optimal** C_P
0.5	1.23	0.75 C_L

0.2	3.29	3.87 C_L
0.1	6.64	14.98 C_L
0.05	13.32	59.43 C_L
0.04	16.66	92.76 C_L
0.03	22.2	164.78 C_L
0.02	33.3	370.54 C_L
0.01	66.7	1481.65 C_L
0.007	95.23	3023.6 C_L

Experiments have shown that it is not unreasonable to achieve a drag ratio (γ) of about 0.01 at a lift coefficient of 0.6. This would give a C_P of about 889. This is substantially better than the best drag-based machine, and explains why lift-based machines are superior.

In the analysis given here, there is an inconsistency compared to typical wind turbine non-dimensionalization. As stated in the preceding section, the *A* (area) in the C_P non-dimensionalization is not always the same as the *A* in the force equations (CL) and (CD). Typically for C_P the A is the area swept by the rotor blade in its motion. For C_L and C_D A is the area of the turbine wing section. For drag based machines, these two areas are almost identical so there is little difference. To make the lift based results comparable to the drag results, the area of the wing section was used to non-dimensionalize power. The results here could be interpreted as power per unit of material. Given that the material represents the cost (wind is free), this is a better variable for comparison.

If one were to apply conventional non-dimensionalization, more information on the motion of the blade would be required. However the discussion on Horizontal Axis Wind Turbines will show that the maximum C_P there is 16/27. Thus, even by conventional non-dimensional analysis lift based machines are superior to drag based machines.

There are several idealizations to the analysis. In any lift-based machine (aircraft included) with finite wings, there is a wake that affects the incoming flow and creates induced drag. This phenomenon exists in wind turbines and was neglected in this analysis. Including induced drag requires information specific to the topology, In these cases it is expected that both the optimal speed-ratio and the optimal C_P would be less. The analysis focused on the aerodynamic potential, but neglected structural aspects. In reality most optimal wind-turbine design becomes a compromise between optimal aerodynamic design and optimal structural design.

Horizontal-axis Wind Turbine

The aerodynamics of a horizontal-axis wind turbine (HAWT) are not straightforward. The air flow at the blades is not the same as the airflow further away from the turbine. The very nature of the way in which energy is extracted from the air also causes air to be deflected by the turbine. In addition, the aerodynamics of a wind turbine at the rotor surface exhibit phenomena rarely seen in other aerodynamic fields.

Axial Momentum and the Lanchester–Betz–Joukowsky limit

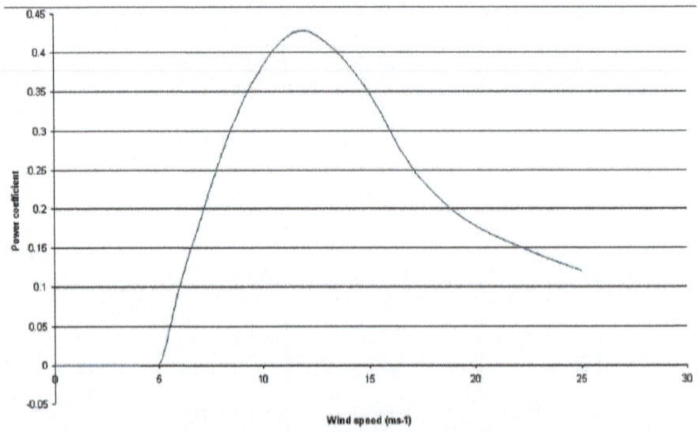

Wind turbine power coefficient

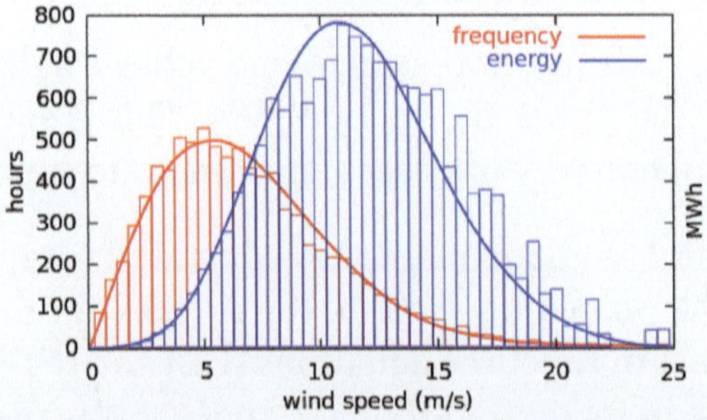

Distribution of wind speed (red) and energy generated (blue). The histogram shows measured data, while the curve is the Rayleigh model distribution for the same average wind speed.

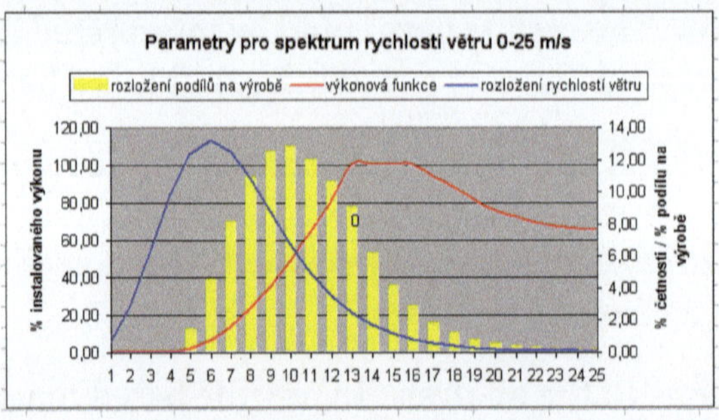

Distribution of wind speed (blue) and energy generated (yellow).

Energy in fluid is contained in four different forms: gravitational potential energy, thermodynamic pressure, kinetic energy from the velocity and finally thermal energy. Gravitational and thermal energy have a negligible effect on the energy extraction process. From a macroscopic point of view, the air flow about the wind turbine is at atmospheric pressure. If pressure is constant then only kinetic energy is extracted. However up close near the rotor itself the air velocity is constant as it passes through the rotor plane. This is because of conservation of mass. The air that passes through the rotor cannot slow down because it needs to stay out of the way of the air behind it. So at the rotor the energy is extracted by a pressure drop. The air directly behind the wind turbine is at sub-atmospheric pressure; the air in front is under greater than atmospheric pressure. It is this high pressure in front of the wind turbine that deflects some of the upstream air around the turbine.

Frederick W. Lanchester was the first to study this phenomenon in application to ship propellers, five years later Nikolai Yegorovich Zhukovsky and Albert Betz independently arrived at the same results. It is believed that each researcher was not aware of the others' work because of World War I and the Bolshevik Revolution. Thus formally, the proceeding limit should be referred to as the Lanchester–Betz–Joukowsky limit. In general Albert Betz is credited for this accomplishment because he published his work in a journal that had a wider circulation, while the other two published it in the publication associated with their respective institution, thus it is widely known as simply the Betz Limit.

This is derived by looking at the axial momentum of the air passing through the wind turbine. As stated above some of the air is deflected away from the turbine. This causes the air passing through the rotor plane to have a smaller velocity than the free stream velocity. The ratio of this reduction to that of the air velocity far away from the wind turbine is called the axial induction factor. It is defined as below:

$$a \equiv \frac{U_1 - U_2}{U_1}$$

where a is the axial induction factor, U_1 is the wind speed far away upstream from the rotor, and U_2 is the wind speed at the rotor.

The first step to deriving the Betz limit is applying conservation of angular momentum. As stated above the wind loses speed after the wind turbine compared to the speed far away from the turbine. This would violate the conservation of momentum if the wind turbine was not applying a thrust force on the flow. This thrust force manifests itself through the pressure drop across the rotor. The front operates at high pressure while the back operates at low pressure. The pressure difference from the front to back causes the thrust force. The momentum lost in the turbine is balanced by the thrust force.

Another equation is needed to relate the pressure difference to the velocity of the flow near the turbine. Here the Bernoulli equation is used between the field flow and the flow near the wind turbine. There is one limitation to the Bernoulli equation: the equation cannot be applied to fluid passing through the wind turbine. Instead conservation of mass is used to relate the incoming air to the outlet air. Betz used these equations and managed to solve the velocities of the flow in the far wake and near the wind turbine in terms of the far field flow and the axial induction factor. The velocities are given below as:

$$\begin{aligned} U_2 &= U_1(1-a) \\ U_4 &= U_1(1-2a) \end{aligned}$$

U_4 is introduced here as the wind velocity in the far wake. This is important because the power extracted from the turbine is defined by the following equation. However the Betz limit is given in terms of the coefficient of power C_p . The coefficient of power is similar to efficiency but not the same. The formula for the coefficient of power is given beneath the formula for power:

$$P = 0.5\rho A U_2 (U_1^2 - U_4^2)$$

$$C_p \equiv \frac{P}{0.5\rho A U_1^3}$$

Betz was able to develop an expression for C_p in terms of the induction factors. This is done by the velocity relations being substituted into power and power is substituted into the coefficient of power definition. The relationship Betz developed is given below:

$$C_p = 4a(1-a)^2$$

The Betz limit is defined by the maximum value that can be given by the above formula. This is found by taking the derivative with respect to the axial induction factor, setting it to zero and solving for the axial induction factor. Betz was able to show that the optimum axial induction factor is one third. The optimum axial induction factor was then used to find the maximum coefficient of power. This maximum coefficient is the Betz limit. Betz was able to show that the maximum coefficient of power of a wind turbine is 16/27. Airflow operating at higher thrust will cause the axial induction factor to rise above the optimum value. Higher thrust cause more air to be deflected away from the turbine. When the axial induction factor falls below the optimum value the wind turbine is not extracting all the energy it can. This reduces pressure around the turbine and allows more air to pass through the turbine, but not enough to account for the lack of energy being extracted.

The derivation of the Betz limit shows a simple analysis of wind turbine aerodynamics. In reality there is a lot more. A more rigorous analysis would include wake rotation, the effect of variable geometry. The effect of airfoils on the flow is a major component of wind turbine aerodynamics. Within airfoils alone, the wind turbine aerodynamicist has to consider the effect of surface roughness, dynamic stall tip losses, solidity, among other problems.

Angular Momentum and Wake Rotation

The wind turbine described by Betz does not actually exist. It is merely an idealized wind turbine described as an actuator disk. It's a disk in space where fluid energy is simply extracted from the air. In the Betz turbine the energy extraction manifests itself through thrust. The equivalent turbine described by Betz would be a horizontal propeller type operating with infinite blades at infinite tip speed ratios and no losses. The tip speed ratio is ratio of the speed of the tip relative to the free stream flow. This turbine is not too far from actual wind turbines. Actual turbines are rotating blades. They typically operate at high tip speed ratios. At high tip speed ratios three blades are sufficient to interact with all the air passing through the rotor plane. Actual turbines still produce considerable thrust forces.

One key difference between actual turbines and the actuator disk, is that the energy is extracted through torque. The wind imparts a torque on the wind turbine, thrust is a necessary by-product of torque. Newtonian physics dictates that for every action there is an equal and opposite reaction. If the wind imparts a torque on the blades then the blades must be imparting a torque on the wind. This torque would then cause the flow to rotate. Thus the flow in the wake has two components, axial and tangential. This tangential flow is referred to as wake rotation.

Torque is necessary for energy extraction. However wake rotation is considered a loss. Accelerating the flow in the tangential direction increases the absolute velocity. This in turn increases the amount of kinetic energy in the near wake. This rotational energy is not dissipated in any form that would allow for a greater pressure drop (Energy extraction). Thus any rotational energy in the wake is energy that is lost and unavailable.

This loss is minimized by allowing the rotor to rotate very quickly. To the observer it may seem like the rotor is not moving fast; however, it is common for the tips to be moving through the air at 6 times the speed of the free stream. Newtonian mechanics defines power as torque multiplied by the rotational speed. The same amount of power can be extracted by allowing the rotor to rotate faster and produce less torque. Less torque means that there is less wake rotation. Less wake rotation means there is more energy available to extract.

Blade Element and Momentum Theory

The simplest model for horizontal axis wind turbine (HAWT) aerodynamics is blade element momentum (BEM) theory. The theory is based on the assumption that the flow at a given annulus does not affect the flow at adjacent annuli. This allows the rotor blade to be analyzed in sections, where the resulting forces are summed over all sections to get the overall forces of the rotor. The theory uses both axial and angular momentum balances to determine the flow and the resulting forces at the blade.

The momentum equations for the far field flow dictate that the thrust and torque will induce a secondary flow in the approaching wind. This in turn affects the flow geometry at the blade. The blade itself is the source of these thrust and torque forces. The force response of the blades is governed by the geometry of the flow, or better known as the angle of attack. Refer to the Airfoil article for more information on how airfoils create lift and drag forces at various angles of attack. This interplay between the far field momentum balances and the local blade forces requires one to solve the momentum equations and the airfoil equations simultaneously. Typically computers and numerical methods are employed to solve these models.

There is a lot of variation between different versions of BEM theory. First, one can consider the effect of wake rotation or not. Second, one can go further and consider the pressure drop induced in wake rotation. Third, the tangential induction factors can be solved with a momentum equation, an energy balance or orthogonal geometric constraint; the latter a result of Biot–Savart law in vortex methods. These all lead to different set of equations that need to be solved. The simplest and most widely used equations are those that consider wake rotation with the momentum equation but ignore the pressure drop from wake rotation. Those equations are given below. a is the axial component of the induced flow, a' is the tangential component of the induced flow. σ is the solidity of the rotor, ϕ is the local inflow angle. C_n and C_t are the coefficient of normal force and the coefficient of tangential force respectively. Both these coefficients are defined with the resulting lift and drag coefficients of the airfoil:

$$a = \frac{1}{\frac{4}{C_n \sigma} \sin^2 \phi + 1}$$

$$a' = \frac{1}{\frac{4}{C_t \sigma} \sin \phi \cos \phi - 1}$$

Corrections to Blade Element Momentum Theory

Blade element momentum (BEM) theory alone fails to represent accurately the true physics of real wind turbines. Two major shortcomings are the effects of a discrete number of blades and far field effects when the turbine is heavily loaded. Secondary shortcomings originate from having to deal with transient effects like dynamic stall, rotational effects like the Coriolis force and centrifugal pumping, and geometric effects that arise from coned and yawed rotors. The current state of the art in BEM uses corrections to deal with these major shortcomings. These corrections are discussed below. There is as yet no accepted treatment for the secondary shortcomings. These areas remain a highly active area of research in wind turbine aerodynamics.

The effect of the discrete number of blades is dealt with by applying the Prandtl tip loss factor. The most common form of this factor is given below where B is the number of blades, R is the outer radius and r is the local radius. The definition of F is based on actuator disk models and not directly applicable to BEM. However the most common application multiplies induced velocity term by F in the momentum equations. As in the momentum equation there are many variations for applying F, some argue that the mass flow should be corrected in either the axial equation, or both axial and tangential equations. Others have suggested a second tip loss term to account for the reduced blade forces at the tip. Shown below are the above momentum equations with the most common application of *F*:

$$F = \frac{2}{\pi}\arccos\left[e^{-\frac{B(R-r)}{2r\sin\phi}}\right]$$

$$a = \frac{1}{\frac{4}{C_n\sigma}F\sin^2\phi + 1}$$

$$a' = \frac{1}{\frac{4}{C_t\sigma}F\sin\phi\cos\phi - 1}$$

The typical momentum theory applied in BEM is only effective for axial induction factors up to 0.4 (thrust coefficient of 0.96). Beyond this point the wake collapses and turbulent mixing occurs. This state is highly transient and largely unpredictable by theoretical means. Accordingly, several empirical relations have been developed. As the usual case there are several version, however a simple one that is commonly used is a linear curve fit given below, with $a_c = 0.2$. The turbulent wake function given excludes the tip loss function, however the tip loss is applied simply by multiplying the resulting axial induction by the tip loss function.

$$C_T = 4\left[a_c^2 + (1-2a_c)a\right] \text{ when } a > a_c$$

The terms C_T and C_t represent different quantities. The first one is the thrust coefficient of the rotor, which is the one which should be corrected for high rotor loading (i.e., for high values of a), while the second one (c_t) is the tangential aerodynamic coefficient of an individual blade element, which is given by the aerodynamic lift and drag coefficients.

Aerodynamic Modeling

BEM is widely used due to its simplicity and overall accuracy, but its originating assumptions limit its use when the rotor disk is yawed, or when other non-axisymmetric effects (like the rotor wake) influence the flow. Limited success at improving predictive accuracy has been made using computational fluid dynamics (CFD) solvers based on Reynolds-averaged Navier–Stokes (RANS) and other similar three-dimensional models such as free vortex methods. These are very computationally intensive simulations to perform for several reasons. First, the solver must accurately model the far-field flow conditions, which can extend several rotor diameters up- and down-stream and include atmospheric boundary layer turbulence, while at the same time resolving the small-scale boundary-layer flow conditions at the blades' surface (necessary to capture blade stall). In addition, many CFD solvers have difficulty meshing parts that move and deform, such as the rotor blades. Finally, there are many dynamic flow phenomena that are not easily modelled by RANS, such as dynamic stall and tower shadow. Due to the computational complexity, it is not currently practical to use these advanced methods for wind turbine design, though research continues in these and other areas related to helicopter and wind turbine aerodynamics.

Free vortex models (FVM) and Lagrangian particle vortex methods (LPVM) are both active areas of research that seek to increase modelling accuracy by accounting for more of the three-dimensional and unsteady flow effects than either BEM or RANS. FVM is similar to lifting line theory in that it assumes that the wind turbine rotor is shedding either a continuous vortex filament from the blade tips (and often the root), or a continuous vortex sheet from the blades' trailing edges. LPVM can use a variety of methods to introduce vorticity into the wake. Biot–Savart summation is used to determine the induced flow field of these wake vorticies' circulations, allowing for better approximations of the local flow over the rotor blades. These methods have largely confirmed much of the applicability of BEM and shed insight into the structure of wind turbine wakes. FVM has limitations due to its origin in potential flow theory, such as not explicitly modelling model viscous behavior (without semi-empirical core models), though LPVM is a fully viscous method. LPVM is more computationally intensive than either FVM or RANS, and FVM still relies on blade element theory for the blade forces.

Wave Power

Wave power is the transport of energy by wind waves, and the capture of that energy to do useful work – for example, electricity generation, water desalination, or the pumping of water (into reservoirs). A machine able to exploit wave power is generally known as a wave energy converter (WEC).

Pelamis Wave Energy Converter on site at the European Marine Energy Centre (EMEC), in 2008

Azura at the US Navy's Wave Energy Test Site (WETS) on Oahu

The mWave converter by Bombora Wave Power

SINN Power Wave Energy Converter

Wave power is distinct from the diurnal flux of tidal power and the steady gyre of ocean currents. Wave-power generation is not currently a widely employed commercial technology, although there have been attempts to use it since at least 1890. In 2008, the first experimental wave farm was opened in Portugal, at the Aguçadoura Wave Park.

Physical Concepts

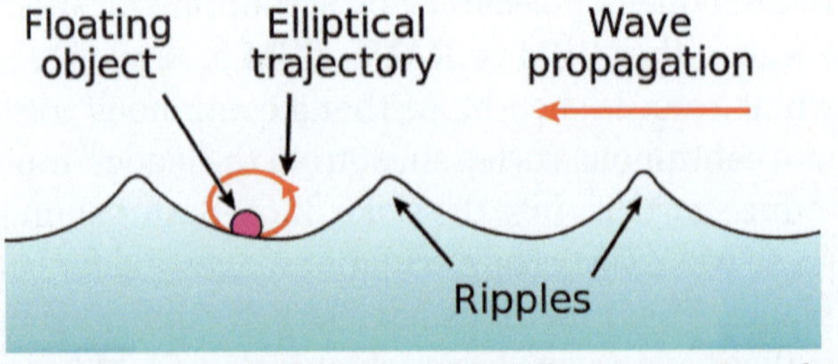

When an object bobs up and down on a ripple in a pond, it follows approximately an elliptical trajectory.

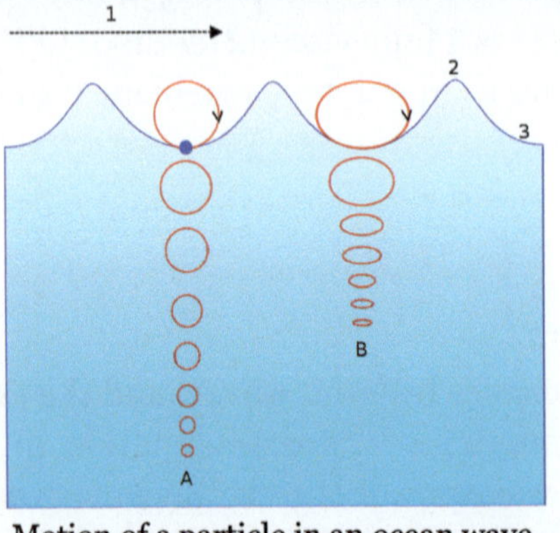

Motion of a particle in an ocean wave.
A = At deep water. The elliptical motion of fluid particles decreases rapidly with increasing depth below the surface. **B** = At shallow water (ocean floor is now at B). The elliptical movement of a fluid particle flattens with decreasing depth. **1** = Propagation direction. **2** = Wave crest. **3** = Wave trough.

Waves are generated by wind passing over the surface of the sea. As long as the waves propagate slower than the wind speed just above the waves, there is an energy transfer from the wind to the waves. Both air pressure differences between the upwind and the lee side of a wave crest, as well as friction on the water surface by the wind, making the water to go into the shear stress causes the growth of the waves.

Wave height is determined by wind speed, the duration of time the wind has been blowing, fetch (the distance over which the wind excites the waves) and by the depth and topography of the sea-floor (which can focus or disperse the energy of the waves). A given wind speed has a matching practical limit over which time or distance will not produce larger waves. When this limit has been reached the sea is said to be "fully developed".

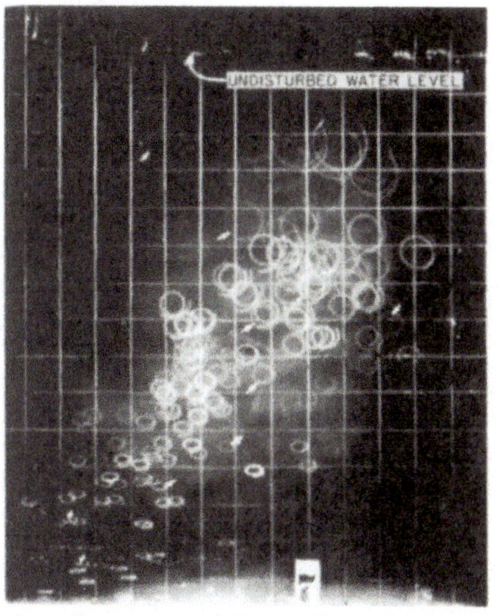

Photograph of the elliptical trajectories of water particles under a – progressive and periodic – surface gravity wave in a wave flume. The wave conditions are: mean water depth d = 2.50 ft (0.76 m), wave height H = 0.339 ft (0.103 m), wavelength λ = 6.42 ft (1.96 m), period T = 1.12 s.

In general, larger waves are more powerful but wave power is also determined by wave speed, wavelength, and water density.

Oscillatory motion is highest at the surface and diminishes exponentially with depth. However, for standing waves (clapotis) near a reflecting coast, wave energy is also present as pressure oscillations at great depth, producing microseisms. These pressure fluctuations at greater depth are too small to be interesting from the point of view of wave power.

The waves propagate on the ocean surface, and the wave energy is also transported horizontally with the group velocity. The mean transport rate of the wave energy through a vertical plane of unit width, parallel to a wave crest, is called the wave energy flux (or wave power, which must not be confused with the actual power generated by a wave power device).

Wave Power Formula

In deep water where the water depth is larger than half the wavelength, the wave energy flux is

$$P = \frac{\rho g^2}{64\pi} H_{m0}^2 T_e \approx \left(0.5 \frac{\text{kW}}{\text{m}^3 \cdot \text{s}}\right) H_{m0}^2\ T_e,$$

with P the wave energy flux per unit of wave-crest length, H_{mo} the significant wave height, T_e the wave energy period, ρ the water density and g the acceleration by gravity. The above formula states that wave power is proportional to the wave energy period and to the square of the wave height. When the significant wave height is given in metres, and the wave period in seconds, the result is the wave power in kilowatts (kW) per metre of wavefront length.

Example: Consider moderate ocean swells, in deep water, a few km off a coastline, with a wave height of 3 m and a wave energy period of 8 seconds. Using the formula to solve for power, we get

$$P \approx 0.5 \frac{\text{kW}}{\text{m}^3 \cdot \text{s}} (3 \cdot \text{m})^2 (8 \cdot \text{s}) \approx 36 \frac{\text{kW}}{\text{m}},$$

meaning there are 36 kilowatts of power potential per meter of wave crest.

In major storms, the largest waves offshore are about 15 meters high and have a period of about 15 seconds. According to the above formula, such waves carry about 1.7 MW of power across each metre of wavefront.

An effective wave power device captures as much as possible of the wave energy flux. As a result, the waves will be of lower height in the region behind the wave power device.

Wave Energy and Wave-energy Flux

In a sea state, the average(mean) energy density per unit area of gravity waves on the water surface is proportional to the wave height squared, according to linear wave theory:

$$E = \frac{1}{16} \rho g H_{m0}^2,$$

where E is the mean wave energy density per unit horizontal area (J/m²), the sum of kinetic and potential energy density per unit horizontal area. The potential energy density is equal to the kinetic energy, both contributing half to the wave energy density E, as can be expected from the equipartition theorem. In ocean waves, surface tension effects are negligible for wavelengths above a few decimetres.

As the waves propagate, their energy is transported. The energy transport velocity is the group velocity. As a result, the wave energy flux, through a vertical plane of unit width perpendicular to the wave propagation direction, is equal to:

$$P = E c_g,$$

with c_g the group velocity (m/s). Due to the dispersion relation for water waves under the action of gravity, the group velocity depends on the wavelength λ, or equivalently, on the wave period T. Further, the dispersion relation is a function of the water depth h. As a result, the group velocity behaves differently in the limits of deep and shallow water, and at intermediate depths:

Properties of gravity waves on the surface of deep water, shallow water and at intermediate depth, according to linear wave theory					
quantity	**symbol**	**units**	**deep water ($h >$ ½ λ)**	**shallow water ($h <$ 0.05 λ)**	**intermediate depth (all λ and h)**
phase velocity	$c_p = \frac{\lambda}{T} = \frac{\omega}{k}$	m / s	$\frac{g}{2\pi} T$	$\sqrt{gh}$	$\sqrt{\frac{g\lambda}{2\pi} \tanh\left(\frac{2\pi h}{\lambda}\right)}$

group velocity	$c_g = c_p^2 \frac{\partial(\lambda / c_p)}{\partial \lambda} = \frac{\partial \omega}{\partial k}$	m / s	$\frac{g}{4\pi}T$	$\sqrt{gh}$	$\frac{1}{2}c_p\left(1+\frac{4\pi h}{\lambda}\frac{1}{\sinh\left(\frac{4\pi h}{\lambda}\right)}\right)$
ratio	$\frac{c_g}{c_p}$	–	$\frac{1}{2}$	1	$\frac{1}{2}\left(1+\frac{4\pi h}{\lambda}\frac{1}{\sinh\left(\frac{4\pi h}{\lambda}\right)}\right)$
wavelength	λ	m	$\frac{g}{2\pi}T^2$	$T\sqrt{gh}$	for given period T, the solution of: $\left(\frac{2\pi}{T}\right)^2 = \frac{2\pi g}{\lambda}\tanh\left(\frac{2\pi h}{\lambda}\right)$
		general			
wave energy density	E	J / m²	$\frac{1}{16}\rho g H_{m0}^2$		
wave energy flux	P	W / m	$E\, c_g$		
angular fre-quency	ω	rad / s	$\frac{2\pi}{T}$		
wavenumber	k	rad / m	$\frac{2\pi}{\lambda}$		

Deep-water Characteristics and Opportunities

Deep water corresponds with a water depth larger than half the wavelength, which is the common situation in the sea and ocean. In deep water, longer-period waves propagate faster and transport their energy faster. The deep-water group velocity is half the phase velocity. In shallow water, for wavelengths larger than about twenty times the water depth, as found quite often near the coast, the group velocity is equal to the phase velocity.

History

The first known patent to use energy from ocean waves dates back to 1799, and was filed in Paris by Girard and his son. An early application of wave power was a device constructed around 1910 by Bochaux-Praceique to light and power his house at Royan, near Bordeaux in France. It appears that this was the first oscillating water-column type of wave-energy device. From 1855 to 1973 there were already 340 patents filed in the UK alone.

Modern scientific pursuit of wave energy was pioneered by Yoshio Masuda's experiments in the 1940s. He tested various concepts of wave-energy devices at sea, with several hundred units used to power navigation lights. Among these was the concept of extracting power from the angular motion at the joints of an articulated raft, which was proposed in the 1950s by Masuda.

A renewed interest in wave energy was motivated by the oil crisis in 1973. A number of university researchers re-examined the potential to generate energy from ocean waves, among whom notably were Stephen Salter from the University of Edinburgh, Kjell Budal and Johannes Falnes from Norwegian Institute of Technology (now merged into Norwegian University of Science and Technology), Michael E. McCormick from U.S. Naval Academy, David Evans from Bristol University, Michael French from University of Lancaster, Nick Newman and C. C. Mei from MIT.

Stephen Salter's 1974 invention became known as Salter's duck or *nodding duck*, although it was officially referred to as the Edinburgh Duck. In small scale controlled tests, the Duck's curved cam-like body can stop 90% of wave motion and can convert 90% of that to electricity giving 81% efficiency.

In the 1980s, as the oil price went down, wave-energy funding was drastically reduced. Nevertheless, a few first-generation prototypes were tested at sea. More recently, following the issue of climate change, there is again a growing interest worldwide for renewable energy, including wave energy.

The world's first marine energy test facility was established in 2003 to kick start the development of a wave and tidal energy industry in the UK. Based in Orkney, Scotland, the European Marine Energy Centre (EMEC) has supported the deployment of more wave and tidal energy devices than at any other single site in the world. EMEC provides a variety of test sites in real sea conditions. Its grid-connected wave test site is situated at Billia Croo, on the western edge of the Orkney mainland, and is subject to the full force of the Atlantic Ocean with seas as high as 19 metres recorded at the site. Wave energy developers currently testing at the centre include Aquamarine Power, Pelamis Wave Power, ScottishPower Renewables and Wello.

Modern Technology

Wave power devices are generally categorized by the method used to capture the energy of the waves, by location and by the power take-off system. Locations are shoreline, nearshore and offshore. Types of power take-off include: hydraulic ram, elastomeric hose pump, pump-to-shore, hydroelectric turbine, air turbine, and linear electrical generator. When evaluating wave energy as a technology type, it is important to distinguish between the four most common approaches: point absorber buoys, surface attenuators, oscillating water columns, and overtopping devices.

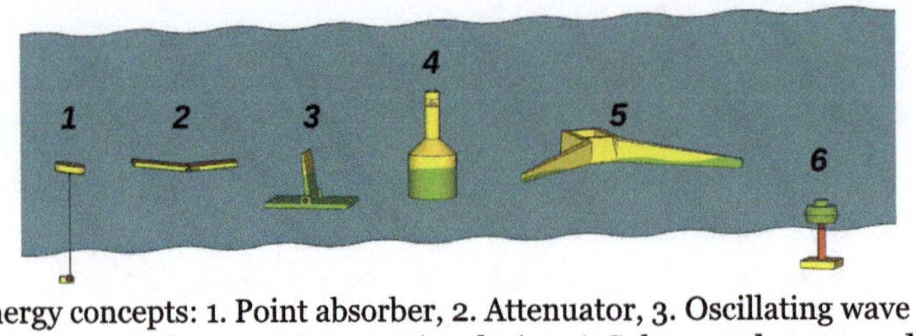

Generic wave energy concepts: 1. Point absorber, 2. Attenuator, 3. Oscillating wave surge converter, 4. Oscillating water column, 5. Overtopping device, 6. Submerged pressure differential

Point Absorber Buoy

This device floats on the surface of the water, held in place by cables connected to the seabed. Buoys use the rise and fall of swells to drive hydraulic pumps and generate electricity. EMF generated by electrical transmission cables and acoustics of these devices may be a concern for marine organisms. The presence of the buoys may affect fish, marine mammals, and birds as potential minor collision risk and roosting sites. Potential also exists for entanglement in mooring lines. Energy removed from the waves may also affect the shoreline, resulting in a recommendation that sites remain a considerable distance from the shore.

Surface Attenuator

These devices act similarly to point absorber buoys, with multiple floating segments connected to one another and are oriented perpendicular to incoming waves. A flexing motion is created by swells that drive hydraulic pumps to generate electricity. Environmental effects are similar to those of point absorber buoys, with an additional concern that organisms could be pinched in the joints.

Oscillating Wave Surge Converter

These devices typically have one end fixed to a structure or the seabed while the other end is free to move. Energy is collected from the relative motion of the body compared to the fixed point. Oscillating wave surge converters often come in the form of floats, flaps, or membranes. Environmental concerns include minor risk of collision, artificial reefing near the fixed point, EMF effects from subsea cables, and energy removal effecting sediment transport. Some of these designs incorporate parabolic reflectors as a means of increasing the wave energy at the point of capture. These capture systems use the rise and fall motion of waves to capture energy. Once the wave energy is captured at a wave source, power must be carried to the point of use or to a connection to the electrical grid by transmission power cables.

Oscillating Water Column

Oscillating Water Column devices can be located on shore or in deeper waters offshore. With an air chamber integrated into the device, swells compress air in the chambers forcing air through an air turbine to create electricity. Significant noise is produced as air is pushed through the turbines, potentially affecting birds and other marine organisms within the vicinity of the device. There is also concern about marine organisms getting trapped or entangled within the air chambers.

Overtopping Device

Overtopping devices are long structures that use wave velocity to fill a reservoir to a greater water level than the surrounding ocean. The potential energy in the reservoir height is then captured with low-head turbines. Devices can be either on shore or floating offshore. Floating devices will have environmental concerns about the mooring system affecting benthic organisms, organisms becoming entangled, or EMF effects produced from subsea cables. There is also some concern regarding low levels of turbine noise and wave energy removal affecting the nearfield habitat.

Submerged Pressure Differential

Submerged pressure differential based converters are a comparatively newer technology utilizing flexible (usually reinforced rubber) membranes to extract wave energy. These converters use the difference in pressure at different locations below a wave to produce a pressure difference within a closed power take-off fluid system. This pressure difference is usually used to produce flow, which drives a turbine and electrical generator. Submerged pressure differential converters frequently use flexible membranes as the working surface between the ocean and the power take-off system. Membranes offer the advantage over rigid structures of being compliant and low mass, which can produce more direct coupling with the wave's energy. Their compliant nature also allows for large changes in the geometry of the working surface, which can be used to tune the response of the converter for specific wave conditions and to protect it from excessive loads in extreme conditions.

A submerged converter may be positioned either on the sea floor or in midwater. In both cases, the converter is protected from water impact loads which can occur at the free surface. Wave loads also diminish in non-linear proportion to the distance below the free surface. This means that by optimizing the depth of submergence for such a converter, a compromise between protection from extreme loads and access to wave energy can be found. Submerged WECs also have the potential to reduce the impact on marine amenity and navigation, as they are not at the surface. Example's of submerged pressure differential converters include M3 Wave and Bombora Wave Power's mWave.

Environmental Effects

Common environmental concerns associated with marine energy developments include:

- The risk of marine mammals and fish being struck by tidal turbine blades;
- The effects of EMF and underwater noise emitted from operating marine energy devices;
- The physical presence of marine energy projects and their potential to alter the behavior of marine mammals, fish, and seabirds with attraction or avoidance;
- The potential effect on nearfield and farfield marine environment and processes such as sediment transport and water quality.

The Tethys database provides access to scientific literature and general information on the potential environmental effects of wave energy.

Potential

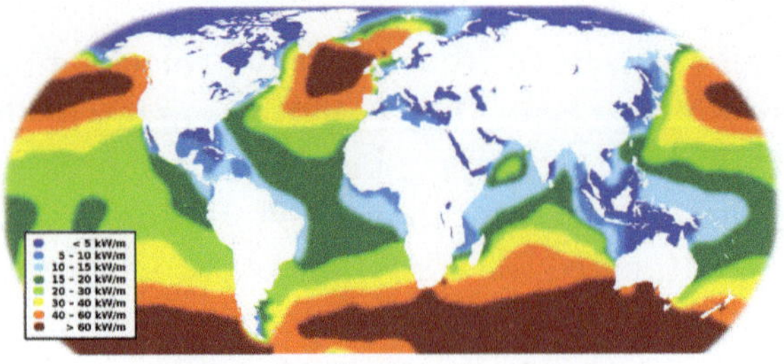

World wave energy resource map

The worldwide resource of wave energy has been estimated to be greater than 2 TW. Locations with the most potential for wave power include the western seaboard of Europe, the northern coast of the UK, and the Pacific coastlines of North and South America, Southern Africa, Australia, and New Zealand. The north and south temperate zones have the best sites for capturing wave power. The prevailing westerlies in these zones blow strongest in winter.

Challenges

There is a potential impact on the marine environment. Noise pollution, for example, could have negative impact if not monitored, although the noise and visible impact of each design vary greatly. Other biophysical impacts (flora and fauna, sediment regimes and water column structure and flows) of scaling up the technology is being studied. In terms of socio-economic challenges, wave farms can result in the displacement of commercial and recreational fishermen from productive fishing grounds, can change the pattern of beach sand nourishment, and may represent hazards to safe navigation. Waves generate about 2,700 gigawatts of power. Of those 2,700 gigawatts, only about 500 gigawatts can be captured with the current technology.

Wave Farms

Portugal

- The Aguçadoura Wave Farm was the world's first wave farm. It was located 5 km (3 mi) offshore near Póvoa de Varzim, north of Porto, Portugal. The farm was designed to use three Pelamis wave energy converters to convert the motion of the ocean surface waves into electricity, totalling to 2.25 MW in total installed capacity. The farm first generated electricity in July 2008 and was officially opened on September 23, 2008, by the Portuguese Minister of Economy. The wave farm was shut down two months after the official opening in November 2008 as a result of the financial collapse of Babcock & Brown due to the global economic crisis. The machines were off-site at this time due to technical problems, and although resolved have not returned to site and were subsequently scrapped in 2011 as the technology had moved on to the P2 variant as supplied to Eon and Scottish Power Renewables. A second phase of the project planned to increase the installed capacity to 21 MW using a further 25 Pelamis machines is in doubt following Babcock's financial collapse.

United Kingdom

- Funding for a 3 MW wave farm in Scotland was announced on February 20, 2007, by the Scottish Executive, at a cost of over 4 million pounds, as part of a £13 million funding package for marine power in Scotland. The first machine was launched in May 2010.
- A facility known as Wave hub has been constructed off the north coast of Cornwall, England, to facilitate wave energy development. The Wave hub will act as giant extension cable, allowing arrays of wave energy generating devices to be connected to the electricity grid. The Wave hub will initially allow 20 MW of capacity to be connected, with potential expansion to 40 MW. Four device manufacturers have so far expressed interest in connecting to the Wave hub. The scientists have calculated that wave energy gathered at Wave Hub will be enough to power up to 7,500 households. The site has the potential to save greenhouse gas emissions of about 300,000 tons of carbon dioxide in the next 25 years.

Australia

- Bombora Wave Power (Bombora Wave Power Pty Ltd) is based in Perth, Western Australia and is currently developing the *mWave* flexible membrane converter. Bombora is currently preparing for a commercial pilot project in Peniche, Portugal.
- A CETO wave farm off the coast of Western Australia has been operating to prove commercial viability and, after preliminary environmental approval, underwent further development. In early 2015 a $100 million, multi megawatt system was connected to the grid, with all the electricity being bought to power HMAS Stirling naval base. Two fully submerged buoys which are anchored to the seabed, transmit the energy from the ocean swell through hydraulic pressure onshore; to drive a generator for electricity, and also to produce fresh water. As of 2015 a third buoy is planned for installation.
- Ocean Power Technologies (OPT Australasia Pty Ltd) is developing a wave farm connected to the grid near Portland, Victoria through a 19 MW wave power station. The project has received an AU $66.46 million grant from the Federal Government of Australia.
- Oceanlinx will deploy a commercial scale demonstrator off the coast of South Australia at Port MacDonnell before the end of 2013. This device, the *greenWAVE*, has a rated electrical capacity of 1MW. This project has been supported by ARENA through the Emerging Renewables Program. The *greenWAVE* device is a bottom standing gravity structure, that does not require anchoring or seabed preparation and with no moving parts below the surface of the water.

United States

- Reedsport, Oregon – a commercial wave park on the west coast of the United States located 2.5 miles offshore near Reedsport, Oregon. The first phase of this project is for ten PB150 PowerBuoys, or 1.5 megawatts. The Reedsport wave farm was scheduled for installation spring 2013. In 2013, the project had ground to a halt because of legal and technical problems.
- Kaneohe Bay Oahu, Hawaii - Navy's Wave Energy Test Site (WETS) currently testing the Azura wave power device The Azura wave power device is 45-ton wave energy converter located at a depth of 30 meters in Kaneohe Bay.

Patents

- WIPO patent application WO2016032360 — 2016 *Pumped-storage system* – "Pressure buffering hydro power" patent application
- U.S. Patent 8,806,865 — 2011 *Ocean wave energy harnessing device* – Pelamis/Salter's Duck Hybrid patent
- U.S. Patent 3,928,967 — 1974 *Apparatus and method of extracting wave energy* – The original "Salter's Duck" patent
- U.S. Patent 4,134,023 — 1977 *Apparatus for use in the extraction of energy from waves on water* – Salter's method for improving "duck" efficiency

- U.S. Patent 6,194,815 — 1999 *Piezoelectric rotary electrical energy generator*
- Wave energy converters utilizing pressure differences US 20040217597 A1 — 2004 *Wave energy converters utilizing pressure differences*

Wave energy is an irregular and oscillating low frequency energy source that can be converted to a 50 Hertz frequency and can then be added to the electric utility grid. Waves get their energy from the wind, which comes from solar energy. Waves gather, store, and transmit this energy thousands of kilometers with very little loss. Though it varies in intensity, it is available twenty four hours a day all round the year. Wave power is renewable, pollution free and environment friendly. Its net potential is better than wind, solar, small hydro or biomass power.

Wave energy technologies rely on the up-and-down motion of waves to generate electricity. There are three basic methods for converting wave energy to electricity.

1. Float or buoy systems that use the rise and fall of ocean swells to drive hydraulic pumps. The object can be mounted to a floating raft or to a device fixed on the ocean bed. A series of anchored buoys rise and fall with the wave. The movement is used to run an electrical generator to produce electricity which is then transmitted ashore by underwater power cables.
2. Oscillating water column devices in which the in-and-out motion of waves at the shore enters a column and force air to turn a turbine. The column fills with water as the wave rises and empties as it descends. In the process, air inside the column is compressed and heats up, creating energy. This energy is harnessed and sent to shore by electrical cable.
3. Tapered channel rely on a shore mounted structure to channel and concentrate the waves driving them into an elevated reservoir. Water flow out of this reservoir is used to generate electricity using standard hydropower technologies.

Wind Wave

In fluid dynamics, wind waves, or wind-generated waves, are surface waves that occur on the free surface of bodies of water (like oceans, seas, lakes, rivers, canals, puddles or ponds). They result from the wind blowing over an area of fluid surface. Waves in the oceans can travel thousands of miles before reaching land. Wind waves on Earth range in size from small ripples, to waves over 100 ft (30 m) high.

Large wave

Ocean waves

When directly generated and affected by local winds, a wind wave system is called a wind sea. After the wind ceases to blow, wind waves are called swells. More generally, a swell consists of wind-generated waves that are not significantly affected by the local wind at that time. They have been generated elsewhere or some time ago. Wind waves in the ocean are called ocean surface waves.

Wind waves have a certain amount of randomness: subsequent waves differ in height, duration, and shape with limited predictability. They can be described as a stochastic process, in combination with the physics governing their generation, growth, propagation and decay—as well as governing the interdependence between flow quantities such as: the water surface movements, flow velocities and water pressure. The key statistics of wind waves (both seas and swells) in evolving sea states can be predicted with wind wave models.

Although waves are usually considered in the water seas of Earth, the hydrocarbon seas of Titan may also have wind-driven waves.

Wave Formation

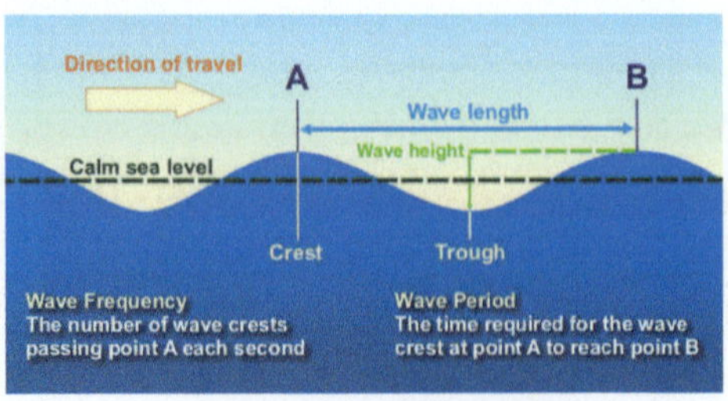

Aspects of a water wave

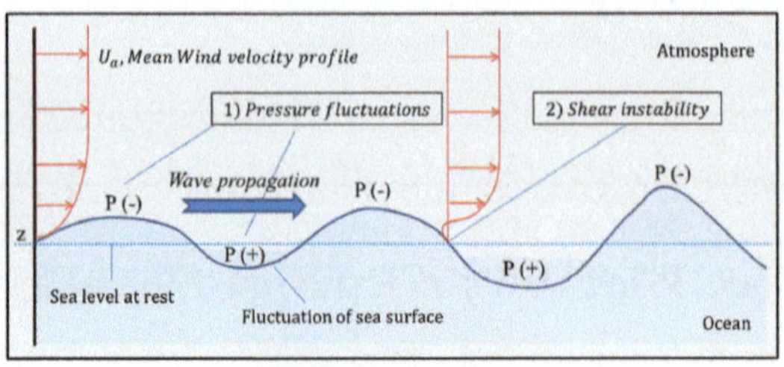

Wave formation

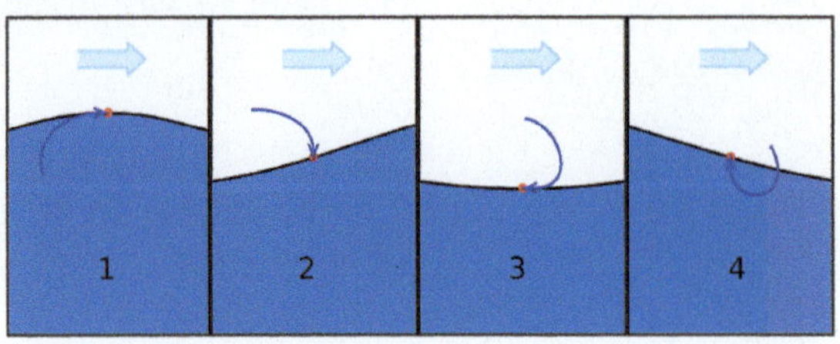

The phases of an ocean surface wave: 1. Wave Crest, where the water masses of the surface layer are moving horizontally in the same direction as the propagating wave front. 2. Falling wave. 3. Trough, where the water masses of the surface layer are moving horizontally in the opposite direction of the wave front direction. 4. Rising wave.

NOAA ship *Delaware II* in bad weather on Georges Bank

The great majority of large breakers seen at a beach result from distant winds. Five factors influence the formation of the flow structures in wind waves:

- Wind speed or strength relative to wave speed—the wind must be moving faster than the wave crest for energy transfer
- The uninterrupted distance of open water over which the wind blows without significant change in direction (called the *fetch*)
- Width of area affected by fetch
- Wind duration — the time for which the wind has blown over the water.
- Water depth

All of these factors work together to determine the size of wind waves and the structure of the flow within them.

The main dimensions associated with waves are:

- Wave height (vertical distance from trough to crest)
- Wave length (distance from crest to crest in the direction of propagation)
- Wave period (time interval between arrival of consecutive crests at a stationary point)
- Wave propagation direction

A fully developed sea has the maximum wave size theoretically possible for a wind of a specific strength, duration, and fetch. Further exposure to that specific wind could only cause a dissipation of energy due to the breaking of wave tops and formation of "whitecaps". Waves in a given area typically have a range of heights. For weather reporting and for scientific analysis of wind wave statistics, their characteristic height over a period of time is usually expressed as *significant wave height*. This figure represents an average height of the highest one-third of the waves in a given time period (usually chosen somewhere in the range from 20 minutes to twelve hours), or in a specific wave or storm system. The significant wave height is also the value a "trained observer" (e.g. from a ship's crew) would estimate from visual observation of a sea state. Given the variability of wave height, the largest individual waves are likely to be somewhat less than twice the reported significant wave height for a particular day or storm.

Wave formation on an initially flat water surface by wind is started by a random distribution of normal pressure of turbulent wind flow over the water. This pressure fluctuation produces normal and tangential stresses in the surface water, which generates waves. It is assumed that:

1. The water is originally at rest.
2. The water is not viscous.
3. The water is irrotational.
4. There is a random distribution of normal pressure to the water surface from the turbulent wind.
5. Correlations between air and water motions are neglected.

The second mechanism involves wind shear forces on the water surface. John W. Miles suggested a surface wave generation mechanism which is initiated by turbulent wind shear flows based on the inviscid Orr-Sommerfeld equation in 1957. He found the energy transfer from wind to water surface is proportional to the curvature of the velocity profile of the wind at the point where the mean wind speed is equal to the wave speed. Since the wind speed profile is logarithmic to the water surface, the curvature has a negative sign at this point. This relation shows the wind flow transferring its kinetic energy to the water surface at their interface.

Assumptions:

1. two-dimensional parallel shear flow
2. incompressible, inviscid water and wind
3. irrotational water
4. slope of the displacement of the water surface is small

Generally these wave formation mechanisms occur together on the water surface and eventually produce fully developed waves.

For example, if we assume a flat sea surface (Beaufort state 0), and a sudden wind flow blows steadily across the sea surface, the physical wave generation process follows the sequence:

1. Turbulent wind forms random pressure fluctuations at the sea surface. Ripples with wavelengths in the order of a few centimetres are generated by the pressure fluctuations. (The Phillips mechanism)
2. The winds keep acting on the initially rippled sea surface causing the waves to become larger. As the waves grow, the pressure differences get larger causing the growth rate to increase. Finally the shear instability expedites the wave growth exponentially. (The Miles mechanism)
3. The interactions between the waves on the surface generate longer waves and the interaction will transfer wave energy from the shorter waves generated by the Miles mechanism to the waves which have slightly lower frequencies than the frequency at the peak wave magnitudes, then finally the waves will be faster than the cross wind speed (Pierson & Moskowitz).

Conditions necessary for a fully developed sea at given wind speeds, and the parameters of the resulting waves					
Wind conditions			**Wave size**		
Wind speed in one direction	**Fetch**	**Wind duration**	**Average height**	**Average wave-length**	**Average period and speed**
19 km/h (12 mph)	19 km (12 mi)	2 hr	0.27 m (0.89 ft)	8.5 m (28 ft)	3.0 sec, 10.2 km/h (9.3 ft/sec)
37 km/h (23 mph)	139 km (86 mi)	10 hr	1.5 m (4.9 ft)	33.8 m (111 ft)	5.7 sec, 21.4 km/h (19.5 ft/sec)
56 km/h (35 mph)	518 km (322 mi)	23 hr	4.1 m (13 ft)	76.5 m (251 ft)	8.6 sec, 32.0 km/h (29.2 ft/sec)
74 km/h (46 mph)	1,313 km (816 mi)	42 hr	8.5 m (28 ft)	136 m (446 ft)	11.4 sec, 42.9 km/h (39.1 ft/sec)
92 km/h (57 mph)	2,627 km (1,632 mi)	69 hr	14.8 m (49 ft)	212.2 m (696 ft)	14.3 sec, 53.4 km/h (48.7 ft/sec)

Types of Wind Waves

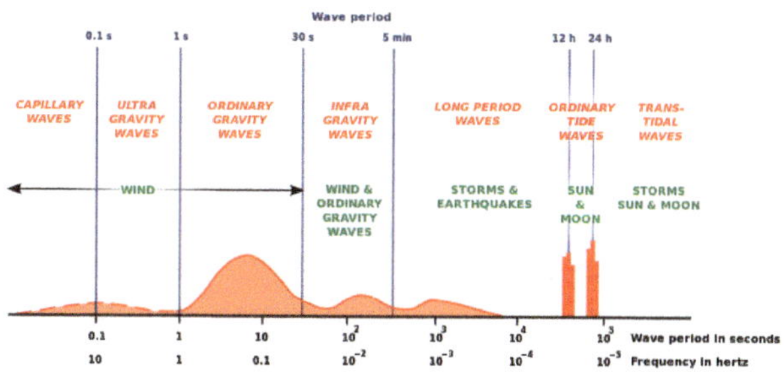

Classification of the spectrum of ocean waves according to wave period

Surf on a rocky irregular bottom. Porto Covo, west coast of Portugal

Three different types of wind waves develop over time:

- Capillary waves, or ripples
- Seas
- Swells

Ripples appear on smooth water when the wind blows, but will die quickly if the wind stops. The restoring force that allows them to propagate is surface tension. Sea waves are larger-scale, often irregular motions that form under sustained winds. These waves tend to last much longer, even after the wind has died, and the restoring force that allows them to propagate is gravity. As waves propagate away from their area of origin, they naturally separate into groups of common direction and wavelength. The sets of waves formed in this way are known as swells.

Individual "rogue waves" (also called "freak waves", "monster waves", "killer waves", and "king waves") much higher than the other waves in the sea state can occur. In the case of the Draupner wave, its 25 m (82 ft) height was 2.2 times the significant wave height. Such waves are distinct from tides, caused by the Moon and Sun's gravitational pull, tsunamis that are caused by underwater earthquakes or landslides, and waves generated by underwater explosions or the fall of meteorites—all having far longer wavelengths than wind waves.

Yet, the largest ever recorded wind waves are common — not rogue — waves in extreme sea states. For example: 29.1 m (95 ft) high waves have been recorded on the RRS Discovery in a sea with 18.5 m (61 ft) significant wave height, so the highest wave is only 1.6 times the significant wave height. The biggest recorded by a buoy (as of 2011) was 32.3 m (106 ft) high during the 2007 typhoon Krosa near Taiwan.

Ocean waves can be classified based on: the disturbing force(s) that create(s) them; the extent to which the disturbing force(s) continue(s) to influence them after formation; the extent to which the restoring force(s) weaken(s) (or flatten) them; and their wavelength or period. Seismic Sea waves have a period of ~20 minutes, and speeds of 760 km/h (470 mph). Wind waves (deep-water waves) have a period of about 20 seconds.

Wave type	Typical wavelength	Disturbing force	Restoring force
Capillary wave	< 2 cm	Wind	Surface tension
Wind wave	60–150 m (200–490 ft)	Wind over ocean	Gravity
Seiche	Large, variable; a function of basin size	Change in atmospheric pressure, storm surge	Gravity
Seismic sea wave (tsunami)	200 km (120 mi)	Faulting of sea floor, volcanic eruption, landslide	Gravity
Tide	Half the circumference of Earth	Gravitational attraction, rotation of Earth	Gravity

The speed of all ocean waves is controlled by gravity, wavelength, and water depth. Most characteristics of ocean waves depend on the relationship between their wavelength and water depth. Wavelength determines the size of the orbits of water molecules within a wave, but water depth determines the shape of the orbits. The paths of water molecules in a wind wave are circular only when the wave is traveling in deep water. A wave cannot "feel" the bottom when it moves through water deeper than half its wavelength because too little wave energy is contained in the small circles below that depth. Waves moving through water deeper than half their wavelength are known as deep-water waves. On the other hand, the orbits of water molecules in waves moving through shallow water are flattened by the proximity of the sea surface bottom. Waves in water shallower than 1/20 their original wavelength are known as shallow-water waves. Transitional waves travel through water deeper than 1/20 their original wavelength but shallower than half their original wavelength.

In general, the longer the wavelength, the faster the wave energy will move through the water. For deep-water waves, this relationship is represented with the following formula:

$$C = L / T$$

where C is speed (celerity), L is wavelength, and T is time, or period (in seconds).

The speed of a deep-water wave may also be approximated by:

$$C = \sqrt{gL / 2\pi}$$

where g is the acceleration due to gravity, 9.8 meters (32 feet) per second squared. Because g and π (3.14) are constants, the equation can be reduced to:

$$C = 1.251\sqrt{L}$$

when C is measured in meters per second and L in meters. Note that in both formulas the wave speed is proportional to the square root of the wavelength.

The speed of shallow-water waves is described by a different equation that may be written as:

$$C = \sqrt{gd} = 3.1\sqrt{d}$$

where C is speed (in meters per second), g is the acceleration due to gravity, and d is the depth of the water (in meters). The period of a wave remains unchanged regardless of the depth of water through which it is moving. As deep-water waves enter the shallows and feel the bottom, however, their speed is reduced and their crests "bunch up," so their wavelength shortens.

Wave Shoaling and Refraction

As waves travel from deep to shallow water, their shape alters (wave height increases, speed decreases, and length decreases as wave orbits become asymmetrical). This process is called shoaling.

Wave refraction is the process by which wave crests realign themselves as a result of decreasing water depths. Varying depths along a wave crest cause the crest to travel at different phase speeds, with those parts of the wave in deeper water moving faster than those in shallow water. This process continues until the crests become (nearly) parallel to the depth contours. Rays—lines normal to wave crests between which a fixed amount of energy flux is contained—converge on local shallows and shoals. Therefore, the wave energy between rays is concentrated as they converge, with a resulting increase in wave height.

Because these effects are related to a spatial variation in the phase speed, and because the phase speed also changes with the ambient current – due to the Doppler shift – the same effects of refraction and altering wave height also occur due to current variations. In the case of meeting an adverse current the wave *steepens*, i.e. its wave height increases while the wave length decreases, similar to the shoaling when the water depth decreases.

Wave Breaking

Big wave breaking

Some waves undergo a phenomenon called "breaking". A breaking wave is one whose base can no longer support its top, causing it to collapse. A wave breaks when it runs into shallow water, or when two wave systems oppose and combine forces. When the slope, or steepness ratio, of a wave is too great, breaking is inevitable.

Individual waves in deep water break when the wave steepness—the ratio of the wave height H to the wavelength λ—exceeds about 0.17, so for $H > 0.17\,\lambda$. In shallow water, with the water depth small compared to the wavelength, the individual waves break when their wave height H is larger than 0.8 times the water depth h, that is $H > 0.8\,h$. Waves can also break if the wind grows strong enough to blow the crest off the base of the wave.

Three main types of breaking waves are identified by surfers or surf lifesavers. Their varying characteristics make them more or less suitable for surfing, and present different dangers.

- Spilling, or rolling: these are the safest waves on which to surf. They can be found in most areas with relatively flat shorelines. They are the most common type of shorebreak
- Plunging, or dumping: these break suddenly and can "dump" swimmers—pushing them to the bottom with great force. These are the preferred waves for experienced surfers. Strong offshore winds and long wave periods can cause dumpers. They are often found where there is a sudden rise in the sea floor, such as a reef or sandbar.
- Surging: these may never actually break as they approach the water's edge, as the water below them is very deep. They tend to form on steep shorelines. These waves can knock swimmers over and drag them back into deeper water.

Physics of Waves

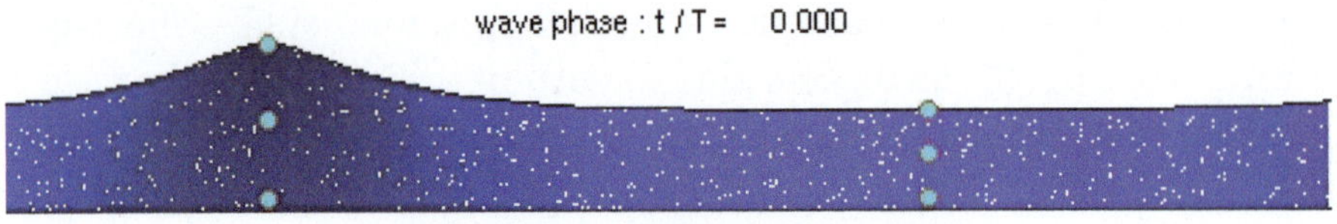

Stokes drift in shallow water waves

Wind waves are mechanical waves that propagate. along the interface between water and air; the restoring force is provided by gravity, and so they are often referred to as surface gravity waves. As the wind blows, pressure and friction perturb the equilibrium of the water surface and transfer energy from the air to the water, forming waves. The initial formation of waves by the wind is described in the theory of Phillips from 1957, and the subsequent growth of the small waves has been modeled by Miles, also in 1957.

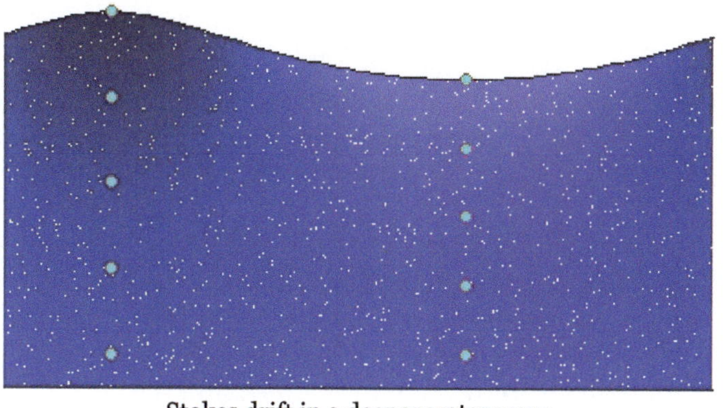

Stokes drift in a deeper water wave

In linear plane waves of one wavelength in deep water, parcels near the surface move not plainly up and down but in circular orbits: forward above and backward below (compared the wave propagation direction). As a result, the surface of the water forms not an exact sine wave, but more a trochoid with the sharper curves upwards—as modeled in trochoidal wave theory. Wind waves are thus a combination of transversal and longitudinal waves.

When waves propagate in shallow water, (where the depth is less than half the wavelength) the particle trajectories are compressed into ellipses.

In reality, for finite values of the wave amplitude (height), the particle paths do not form closed orbits; rather, after the passage of each crest, particles are displaced slightly from their previous positions, a phenomenon known as Stokes drift.

As the depth below the free surface increases, the radius of the circular motion decreases. At a depth equal to half the wavelength λ, the orbital movement has decayed to less than 5% of its value at the surface. The phase speed (also called the celerity) of a surface gravity wave is – for pure periodic wave motion of small-amplitude waves – well approximated by

$$c = \sqrt{\frac{g\lambda}{2\pi} \tanh\left(\frac{2\pi d}{\lambda}\right)}$$

where

c = phase speed;

λ = wavelength;

d = water depth;

g = acceleration due to gravity at the Earth's surface.

In deep water, where $d \geq \frac{1}{2}\lambda$, so $\frac{2\pi d}{\lambda} \geq \pi$ and the hyperbolic tangent approaches 1, the speed c approximates

$$c_{\text{deep}} = \sqrt{\frac{g\lambda}{2\pi}}.$$

In SI units, with c_{deep} in m/s, $c_{\text{deep}} \approx 1.25\sqrt{\lambda}$, when λ is measured in metres. This expression tells us that waves of different wavelengths travel at different speeds. The fastest waves in a storm are the ones with the longest wavelength. As a result, after a storm, the first waves to arrive on the coast are the long-wavelength swells.

For intermediate and shallow water, the Boussinesq equations are applicable, combining frequency dispersion and nonlinear effects. And in very shallow water, the shallow water equations can be used.

If the wavelength is very long compared to the water depth, the phase speed (by taking the limit of c when the wavelength approaches infinity) can be approximated by

$$c_{\text{shallow}} = \lim_{\lambda \to \infty} c = \sqrt{gd}.$$

On the other hand, for very short wavelengths, surface tension plays an important role and the phase speed of these gravity-capillary waves can (in deep water) be approximated by

$$c_{\text{gravity-capillary}} = \sqrt{\frac{g\lambda}{2\pi} + \frac{2\pi S}{\rho\lambda}}$$

where

S = surface tension of the air-water interface;

ρ = density of the water.

When several wave trains are present, as is always the case in nature, the waves form groups. In deep water the groups travel at a group velocity which is half of the phase speed. Following a single wave in a group one can see the wave appearing at the back of the group, growing and finally disappearing at the front of the group.

As the water depth d decreases towards the coast, this will have an effect: wave height changes due to wave shoaling and refraction. As the wave height increases, the wave may become unstable when the crest of the wave moves faster than the trough. This causes *surf*, a breaking of the waves.

The movement of wind waves can be captured by wave energy devices. The energy density (per unit area) of regular sinusoidal waves depends on the water density ρ, gravity acceleration g and the wave height H (which, for regular waves, is equal to twice the amplitude, a):

$$E = \frac{1}{8}\rho g H^2 = \frac{1}{2}\rho g a^2$$

The velocity of propagation of this energy is the group velocity.

Wind Wave Models

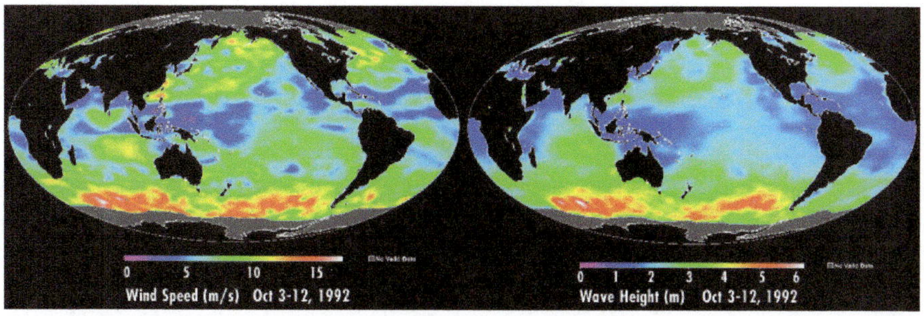

The image shows the global distribution of wind speed and wave height as observed by NASA's TOPEX/Poseidon's dual-frequency radar altimeter from October 3 to October 12, 1992.

Simultaneous observations of wind speed and wave height are helping scientists to predict ocean waves. Wind speed is determined by the strength of the radar signal after it has bounced off the ocean surface and returned to the satellite. A calm sea serves as a good reflector and returns a strong signal; a rough sea tends to scatter the signals and returns a weak pulse. Wave height is determined by the shape of the return radar pulse. A calm sea with low waves returns a condensed pulse whereas a rough sea with high waves returns a stretched pulse. Comparing the two images above shows a high degree of correlation between wind speed and wave height. The strongest winds (33.6 mph; 54.1 km/h) and highest waves are found in the Southern Ocean. The weakest winds — shown as areas of magenta and dark blue — are generally found in the tropical oceans.

Surfers are very interested in the wave forecasts. There are many websites that provide predictions of the surf quality for the upcoming days and weeks. Wind wave models are driven by more general weather models that predict the winds and pressures over the oceans, seas and lakes.

Wind wave models are also an important part of examining the impact of shore protection and beach nourishment proposals. For many beach areas there is only patchy information about the wave climate, therefore estimating the effect of wind waves is important for managing littoral environments.

Seismic Signals

Ocean water waves generate land seismic waves that propagate hundreds of kilometers into the land. These seismic signals usually have the period of 6 ± 2 seconds. Such recordings were first reported and understood in about 1900.

There are two types of seismic "ocean waves". The primary waves are generated in shallow waters by direct water wave-land interaction and have the same period as the water waves (10 to

16 seconds). The more powerful secondary waves are generated by the superposition of ocean waves of equal period traveling in opposite directions, thus generating standing gravity waves – with an associated pressure oscillation at half the period, which is not diminishing with depth. The theory for microseism generation by standing waves was provided by Michael Longuet-Higgins in 1950, after in 1941 Pierre Bernard suggested this relation with standing waves on the basis of observations.

Internal Waves

Internal waves can form at the boundary between water layers of different densities. These sub-surface waves are called internal waves. As is the case with ocean waves at the air-ocean interface, internal waves possess troughs, crests, wavelength, and period. Internal waves move very slowly because the density difference between the joined media is very small. Internal waves occur in the ocean at the base of the pycnocline, especially at the bottom edge of a steep thermocline. The wave height of internal waves may be greater than 30 meters (98 feet), causing the pycnocline to undulate slowly through a considerable depth. Their wavelength often exceeds 0.8 kilometres (0.50 mi) and their periods are typically 5 to 8 minutes. Internal waves are generated by wind energy, tidal energy, and ocean currents. Surface manifestations of internal waves have been photographed from space.

Internal waves may mix nutrients into surface water and trigger plankton blooms. They can also affect submarines and oil platforms.

Estimation of Power Associated with Waves

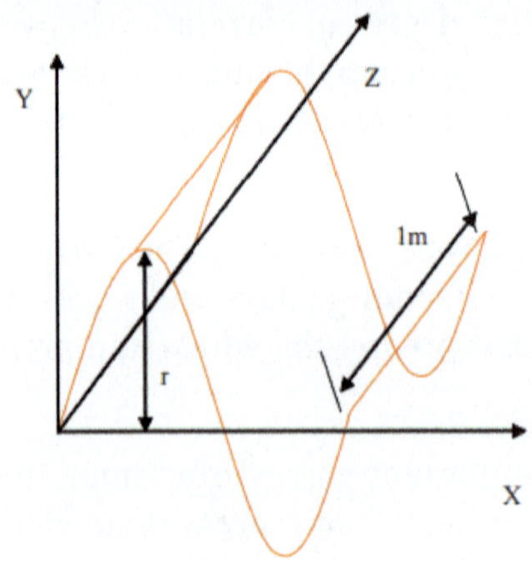

A Wave of unit Width and amplitude r moving along x-direction

Let the ampitude of the wave at the surface of the wave be 'a'.

Wave is traveling in the x-direction. Wave amplitude is in Y direction.

$r = ae^{ky}$

Where r is amplitude of wave at distance y in the negative direction.

It means that the amplitude of oscillations of particles decreases as we move in to the depth of the sea.

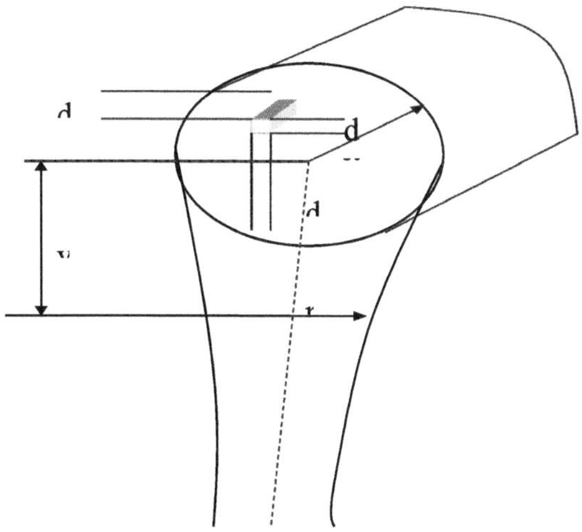

dv=dx.dy.dz

dx*dy is the volume per unit width of the wave.

Ek = kinetic energy per unit length in X direction.

dEk*dx= K.E of particle of width dx

$$
\begin{aligned}
&= (1/2)mv^2 \\
&= (1/2).dx.dy.\rho.V^2 \\
&= (1/2).dx.dy.\rho.(\omega.r)^2
\end{aligned}
$$

$$
\begin{aligned}
dEk &= (1/2).dy.\rho.(\omega.r)^2 \\
&= (1/2).\rho.\omega^{2\cdot}a^{2\cdot}e^{2y}.dy
\end{aligned}
$$

$$
\begin{aligned}
J^0{}_{-\infty} dEk &= (1/2).\rho.\omega^2.a^2.J^0{}_{-\infty}e^{2y}.dy \\
&= \rho^*\omega^2 * \mathrm{a}^2/(4*K)
\end{aligned}
$$

$$K = 2\pi/\lambda$$

$$\lambda = 2\pi g/\omega^2$$

$$Ek = (1/4).\rho.a^2.g$$

$$Ep = (1/4).\rho.a^2.g$$

$$Total\ wave\ energy = (1/2)^*.\rho.a^2.g$$

$$E\lambda = (1/4)^*.\rho.a^2.g.\lambda$$

$$= (1/4\pi).\rho.a^2.\mathrm{T}^2$$

Power associated with the wave per unit width, P/unit width $= (1/8\pi)\rho a^2 g^2 T$

Water waves can be considered to travel along the surface of the sea with an approximate sinusoidal profile. They can be characterized in terms of the distance between successive crests (the wavelength, λ) and the time between successive crests (the period, T). In deep water these parameters are related as follows:

$$\lambda = gT^2 / 2\pi$$

Where g is the acceleration due to gravity.

The velocity of the waves, C, is given by the following relationship:

$$C = \lambda / T$$

Hence, longer waves travel faster than shorter ones. This effect is seen in hurricane areas, where long waves generally travel faster than the storm generating them and so the arrival of the hurricane is often preceded by heavy surf on the coast.

The power (P) in such waves can also be described by use of these parameters and the wave height, H:

$$P = \rho g^2 TH^2 / 32\pi$$

Where ρ is the density of seawater and P is expressed per unit crest length of the wave. Most of the energy within a wave is contained near the surface and falls off sharply with depth. Therefore, most wave energy devices are designed to float (or in the case of bottom standing devices to be in shallow water) and so pierce the water surface in order to maximize the energy available for capture.

However, in practice waves are far from ideal. In nature, waves are irregular and can be described by statistical models. If the wave conditions are measured, over 20 minutes for example, the mean wave height Hm and the significant wave height Hs can be calculated. The significant wave height is defined as the average of the highest 33% of the waves. Under such circumstances, the wave power can be stated to be:

$$P = 0.55\ H2s\ Tz\ kW / m \text{ length of wave crest}$$

Where Tz is the zero crossing period [Charlier, Justus 1993].

Problems Associated with Utilization of Wave Energy

- The first problem is shortage of wave data, especially close to the shore. The required information includes the record of wave height with time and directional properties of the wave.
- The second problem relates to effect of the wave forces on the structures of the devices. A device, which absorbs or reflects wave energy must also absorb or reflect momentum and must be subjected to mean horizontal force. This force must be balanced by an opposing force provided by fixing the device to seabed or by arranging for opposing forces to be exerted by the sea on other parts of the devices. Fixing the seabed by the cables obviously introduces anchoring and mooring problems, whilst forces, which are

not balanced, provide a bending moment on the structure, which may be difficult to accommodate.

- The third problem concerns the power generation from slow irregular motion of waves to the steady high-speed rotary motion normally required for electrical machines. The transmission of this power to ashore may also be difficult if flexible cables are required.
- The fourth problem leads to effects of devices on the environment. Changes in the wave climate caused by wave energy converters could affect the local shoreline. Fish feeding and spawning could be affected by these devices, which would also present some hazard to fishing.

Wave Energy Collecting Devices

Shoreline Devices

- Oscillating Water Column (OWC) Devices
- Tapered Channel Devices (TAPCHAN)
- The Pendulor Device

Offshore Devices

- Float-Based Devices
- Moving body Devices

Shoreline Devices

These devices are fixed to or embedded in the shoreline itself, which has the advantage of easier maintenance and/or installation. In addition these would not require deep-water moorings or long lengths of underwater electrical cable. However, they would experience a much less powerful wave regime. This could be partially compensated by natural energy concentration. The deployment of such schemes could be limited by requirements for shoreline geology, tidal range, preservation of coastal scenery, etc.

Oscillating Water Column (OWC) Devices

One major class of shoreline device is the oscillating water column partly submerged concrete or steel structure, which has an opening to the sea below the water line, thereby enclosing a column of air above a column of water. As waves impinge on the device, they cause the water column to rise and fall, which alternately compresses and depressurizes the air column. This air is allowed to flow to and from the atmosphere through a turbine, which drives an electric generator. Both conventional (i.e. unidirectional) and self- rectifying air turbines have been proposed. The axial-flow Wells turbine, invented in the 1970s, is the best-known turbine for this kind of application and has the advantage of not requiring rectifying air valves. A number of OWC devices have been installed world wide, with several of them being built into a breakwater to lower overall construction costs.

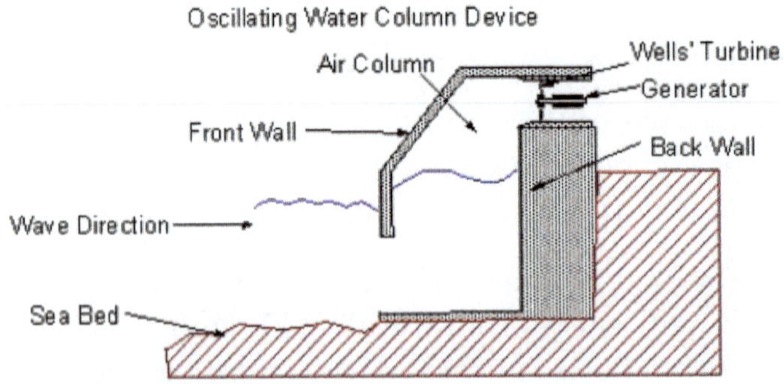

Tapered Channel Devices (Tapchan)

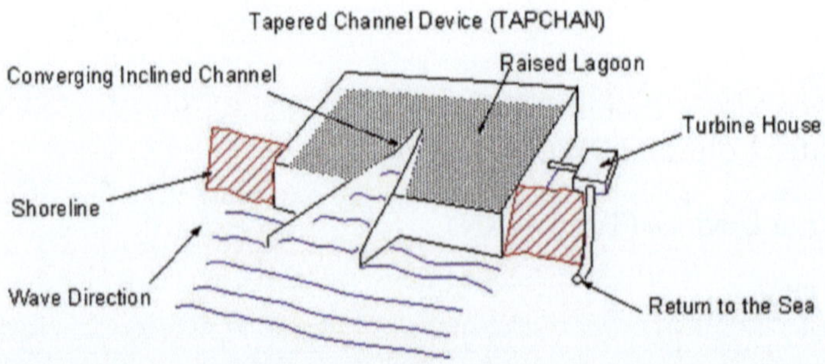

The TAPCHAN comprises a gradually narrowing channel with wall heights typically 3 to 5 m above mean water level. Waves enter the wide end of the channel and, as they propagate down the narrowing channel, the wave height is amplified until the wave crests spill over the walls to the reservoir, which is raised above sea level. The water in the reservoir returns to the sea via a conventional low head turbine, which generates a stable output due to the storage effects of the reservoir. A demonstration device with rated output of 350 kW began operating in 1985 at Toftesfallen, in Norway. The device functioned successfully until the early 1990s, when work on modifying the device destroyed the tapered channel. The potential market for such a device is limited within Europe, because the design requires a small tidal range.

The Pendulor Device

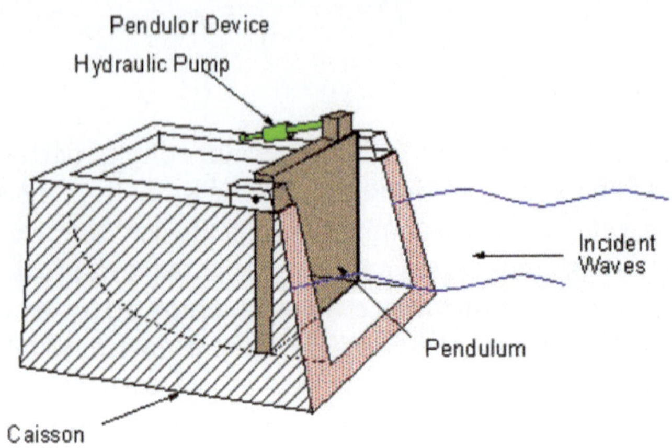

A 5 kW Pendulor test device has been operating in Hokkaido since 1983. It consists of a rectangular box, which is open to the sea at one end. A pendulum flap is hinged over this opening, so that the actions of the waves cause it to swing back and forth. This motion is then used to power a hydraulic pump and generator.

Offshore Devices

This class of device exploits the more powerful wave regimes available in deep water (> 40 m depth) before energy dissipation mechanisms have had a significant effect. In order to extract energy from the waves, the devices need to be at or near the surface (i.e. floating) and so they require flexible moorings and electrical transmission cables. There are many different types of offshore device, some of which are:

Float-based Devices

The simplest concepts extract energy from the vertical motion of a float as it rises and falls with each wave. If the motion of the float is reacted against an anchor or other structure that resists motion, then energy can be extracted. In the Danish Wave Power (DWP) device, this is achieved by anchoring the float to a pump and generator mounted in a concrete box on the seabed (DWP, 1996; Nielsen et al, 1995). Following developmental work, a 1 kW test device was installed near the harbor of Hanstholm, Denmark. This incorporated an air reservoir, which acted as an energy storage system, thereby smoothing the device output. After some initial difficulties, this device performed continuously for several months providing considerable amounts of information.

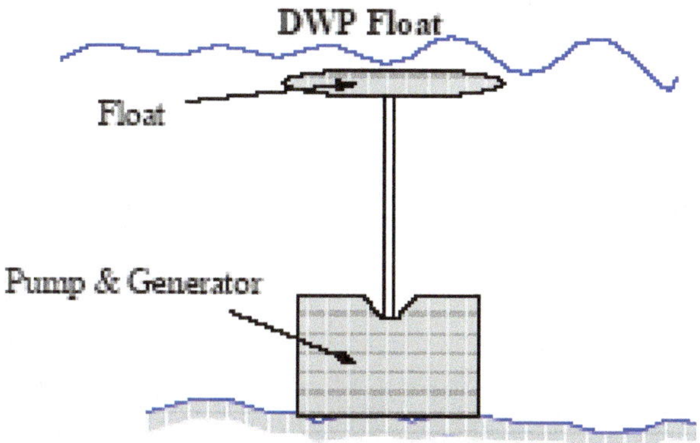

Another example of a device using a float is based on the Hose pump, which is a specially reinforced elastomeric hose whose volume decreases as it is stretched. The interior of the Hose pump is filled with seawater, which is pressurized as the float rises. By using a non-return valve, the device can supply pressurized seawater to a line connecting several Hose pump modules together. This line supplies seawater to a conventional Pelton turbine at pressures between 1 and 4 MPa). Laboratory testing of the Hose pumps was followed by the installation of a single, small-scale module in Lake Lygnern. Later a larger system, comprising five modules connected in parallel to a single turbine and generator, was installed in Lake Lygnern. Despite loss of early systems in storms, a costing exercise was carried out on a 64 MW station comprising 360 modules for emplacement off the Norwegian coast (GES, 1984). The performance of the system was good enough to encourage recent interest in commercial exploitation in using Hose pumps to power navigation buoys.

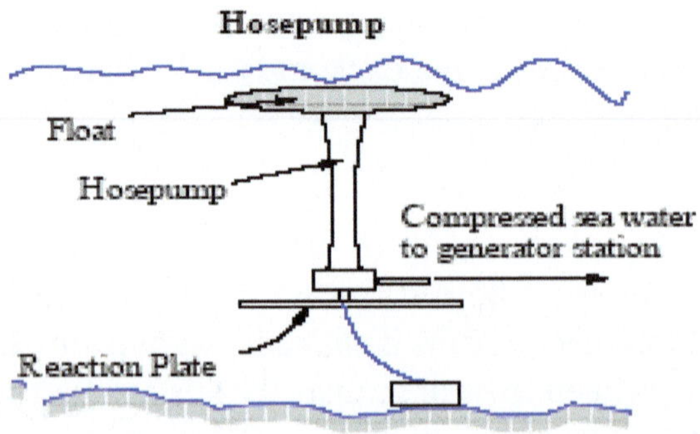

The Pelamis. The Pelamis device is a semi-submerged, articulated structure composed of cylindrical sections linked by hinged joints. The wave-induced motion of these joints is resisted by hydraulic rams, which pump high-pressure oil through hydraulic motors via smoothing accumulators. The hydraulic motors drive electrical generators to produce electricity. A novel joint configuration is used to induce a tuneable, cross-coupled resonant response, which greatly increases power capture in small seas and control of the restraint applied to the joints allows tuning to the particular sea state.

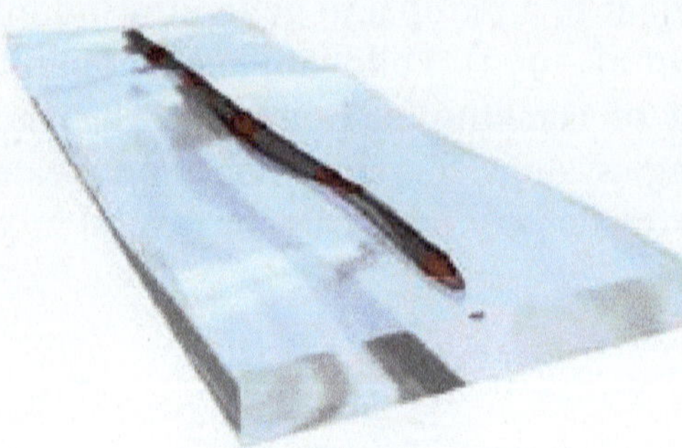

Pelamis drawing

The Archimedes Wave Swing. This consists of a cylindrical, air filled chamber (the "Floater"), which can move vertically with respect to the cylindrical "Basement", which is fixed to the seabed. The air within the 10m – 20m diameter Floater ensures buoyancy. However, a wave passing over the top of the device alternatively pressurizes and depressurizes the air within the Floater, changing this buoyancy. This causes the Floater to move up and down with respect to the Basement and it is this relative motion that is used to produce energy. The design for a 2 MW Pilot scheme is currently being finalized.

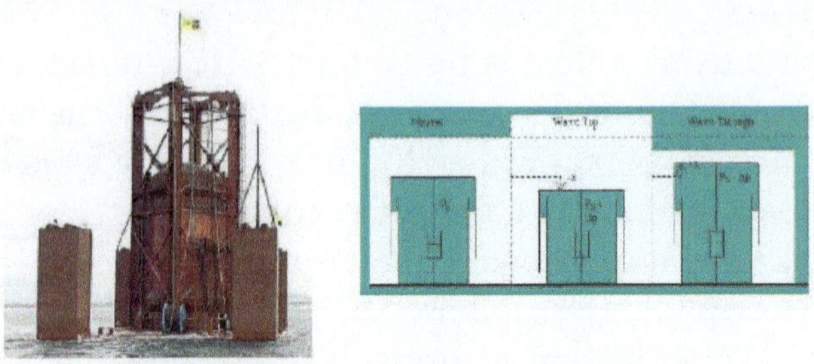

AWS Above-surface and Technical Principle

The status of these designs is slightly less advanced than that for OWCs. However, several devices have been tested in the sea at near full size and so are ready for commercial demonstration. All these devices will continue to benefit from ongoing R&D.

The McCabe Wave Pump. This device consists of three narrow (4 m wide) rectangular steel pontoons, which are hinged together across their beam. These pontoons are aligned so that their longitudinal direction heads into the incoming waves and they can move relative to each other in the waves. The essential aspect of the scheme is the damper plate attached to the central pontoon this increases the inertia of the central pontoon (by effectively adding mass), ensuring that it stays relative still. Therefore, the fore and aft pontoons move in relation to the central pontoon by pitching about the hinges. Energy is extracted from the rotation about the hinge points by linear hydraulic rams mounted between the central and two outer pontoons near the hinges. Control of the characteristics of the hydraulic system allows the device to be tuned to the prevailing sea state and so optimize energy capture. This energy can be used in two ways:

- To provide electricity by driving an hydraulic motor attached to a generator, with a rating of ~ 400 kW
- To produce potable water by supplying pressurized seawater to a reverse osmosis plant. A 40 m long prototype of this device was deployed off the coast of Kilbaha, County Clare, Ireland and a commercial demonstration scheme is currently being built.

Outline of the McCabe Wave Pump

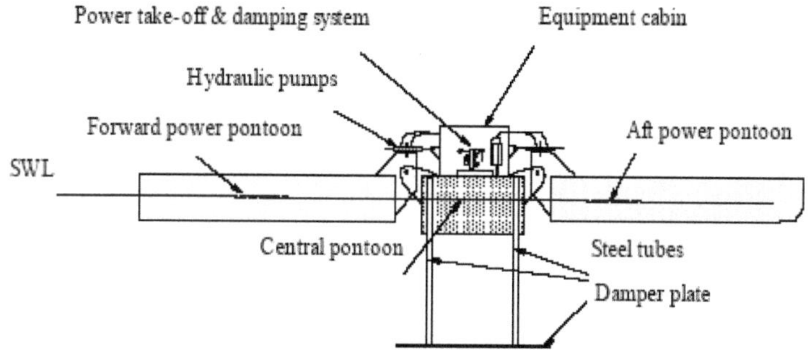

Mooring Arrangement for the McCabe Wave Pump

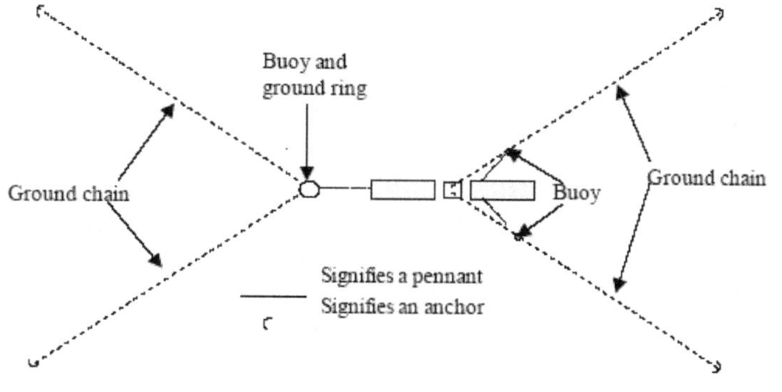

There are some experimental data from small-scale tests, together with a short period of observations on the prototype. These indicate that the MWP can have capture efficiencies >100%. In order to evaluate the performance of the MWP in a range of sea states, its behavior has to be evaluated theoretically. With different wave energy periods (T_e) and the resulting capture efficiency is shown in Figure. This shows a slight reduction in capture efficiency at short wave periods, when compared to the capture efficiency in regular seas.

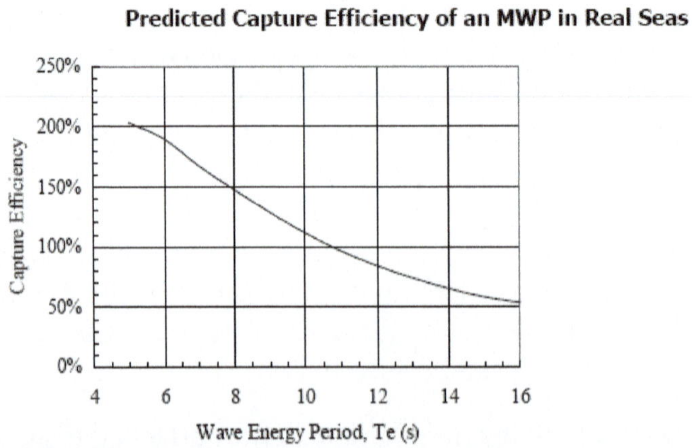

The above predictions apply only to the fore pontoon. However, the aft pontoon will also capture wave energy. It is difficult to determine this theoretically but observations made on the prototype suggest that the aft pontoon picks up ~60% of the energy captured by the fore pontoon. The average capture efficiency was approximately 150%.

Characteristics of the MWP

Device Width	4	m
Sea Power Level	53	kW/m
Capture Efficiency	150	%
Hydraulic efficiency	98	%
Efficiency of hydraulic motor	95	%
Efficiency of generator	96	%
Availability	90	%
Average Output per Device	257	KW
Annual output of scheme	2.25	GWh

Shoreline Devices

Limpet (Land Installed Marine Powered Energy Transformer)

The Limpet a 500KW device is a shoreline-based oscillating water column (OWC), which has been developed by Queen's University of Belfast and Wavegen of Inverness. Wavegen installed a Limpet on the Islay Island in 2000.

Limpet follows the designer gully concept, in which the device is constructed and fixed in place close to the shoreline, being protected from the sea by a rock bund. When the device is completely installed, the bund is removed, allowing the sea access to the device. The device consists of three water columns placed side by side in a man-made recess, which forms a slipway at an angle to the horizontal. In the current design for the island of Islay, the water column boxes are made from a steel-concrete-steel sandwich called BISTEEL, giving a device width of ~21 m. The device is anchored to the rock promontories on either side and to the base. Wave tank tests have shown that the inclined slope increases the capture efficiency. There are two low solidity, counter rotating Wells' turbines each rated at 500 kW. The turbines are placed behind the OWC chambers and the associated electrical equipment is located behind the turbines, where it is protected from seawater splashes by a rock bund. Each set of turbines is protected by a sluice gate, which can prevent the turbines being subjected to green water in stormy seas. The turbines also have flywheels to smooth out energy supply, as well as blow out valves.

Outline Side View of the Limpet

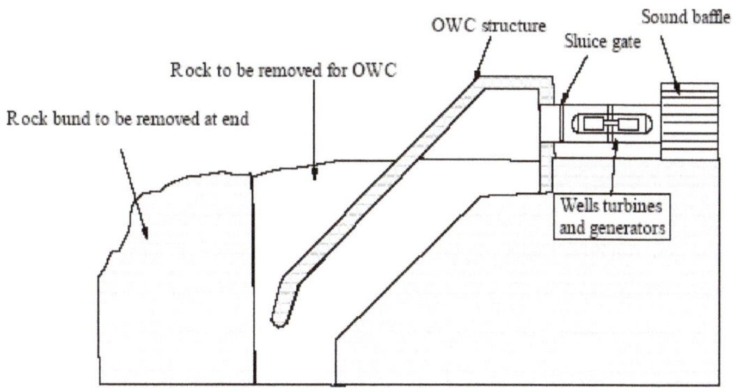

Technical Considerations

- The capture efficiency of the system has been proven in model testing but it is likely to be site-specific.
- The turbines are situated close to the OWC end of the air vents (~ twice the turbine diameter), which could result in non-streamlined airflow and consequential loss of efficiency. However, Wavegen have studied this both theoretically and experimentally and consider it to have only a very small effect.
- The device has a design life of +30 years for the structure but that of the mechanical and electrical plant will be somewhat less. The approach has been to use robust and reliable components, so they should achieve the adequate reliability found when used in other, similar situations.

Replication Potential: The unit size will be ~ 1 MW allowing the system to integrate with nearly all parts of the grid. Wavegen have produced a modular system, tailored for deployment on islands and in remote locations. Therefore, the only limits on its replicability will be economic, geographic and environmental.

Geographic: the scheme needs the right combination of shoreline topography and geography, together with low tidal ranges and closeness to the grid.

Environmental: the original Islay scheme impinged on the shoreline and had some visual impact. Wavegen have tried to address this in developing a low profile, composite roof, modular system with a much-reduced visual impact. They have also considered decommissioning of the device after service and have set aside monies to account for this.

Economic

Capital Cost

This is site-specific and subject to change following ongoing developments. There was a large degree of uncertainty in estimating these costs, because of the lack of detailed information, leading to predictions in the range £ 850,000 to £ 1,160,000.

Operational and Maintenance Costs

The annual operation and maintenance costs were estimated to be approximately £ 23,000. The system is being developed with a view to it being maintained in remote locations by indigenous staff.

Electrical Output

The average wave power level at the Islay site is 20 kW/m. At this location, the device is estimated to have an average electrical output of 206 kW, amounting to 1,800 MWh/year. Details concerning certain aspects of the electrical output are commercially sensitive (e.g. capture efficiency of this sloped OWC chamber). In addition, the energy output from this particular scheme is enhanced by the shape of the cliff, which forms sloping harbor walls either side of the OWC chamber, thereby increasing the effective capture width of the device.

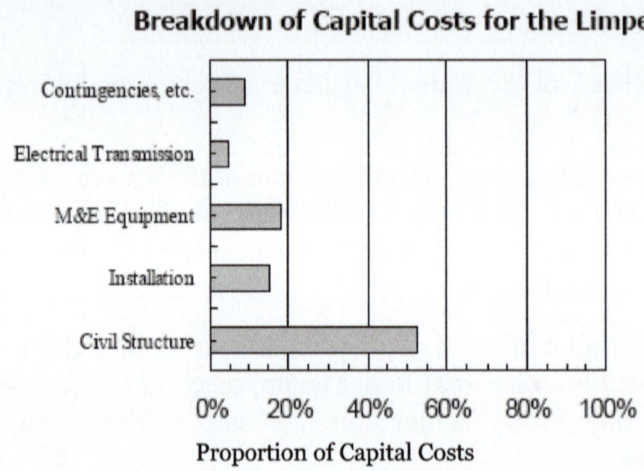

References

- Archer, C. L.; Jacobson, M. Z. (2007). "Supplying Baseload Power and Reducing Transmission Requirements by Interconnecting Wind Farms" (PDF). Journal of Applied Meteorology and Climatology. American Meteorological Society. 46 (11): 1701–1717. Bibcode:2007JApMC..46.1701A. doi:10.1175/2007JAMC1538.1
- History of Wind Energy in Cutler J. Cleveland,(ed) Encyclopedia of Energy Vol.6, Elsevier, ISBN 978-1-60119-433-6, 2007, pp. 421–422
- Fthenakis, V.; Kim, H. C. (2009). "Land use and electricity generation: A life-cycle analysis". Renewable and Sustainable Energy Reviews. 13 (6–7): 1465. doi:10.1016/j.rser.2008.09.017
- Shackleton, Jonathan. "World First for Scotland Gives Engineering Student a History Lesson". The Robert Gordon University. Archived from the original on 17 December 2008. Retrieved 20 November 2008
- Haapala, Karl R.; Prempreeda, Preedanood (2014). "Comparative life cycle assessment of 2.0 MW wind tur-

bines". International Journal of Sustainable Manufacturing. 3 (2): 170. doi:10.1504/IJSM.2014.062496

- Patel, Mukund R. (2006). Wind and Solar Power Systems – Design, analysis and Operation (PDF) (2nd ed.). CRC Press. p. 303. ISBN 0-8493-1570-0
- Gipe, Paul (1993). "The Wind Industry's Experience with Aesthetic Criticism". Leonardo. 26 (3): 243–248. JSTOR 1575818. doi:10.2307/1575818
- "World's second largest offshore wind farm opens off coast of Wales". Wales Online. Archived from the original on 19 June 2015. Retrieved 18 June 2015
- Savenkov, M (2009). "On the truncated weibull distribution and its usefulness in evaluating potential wind (or wave) energy sites" (PDF). University Journal of Engineering and Technology. 1 (1): 21–25. Archived from the original on 22 February 2015
- Wind Turbines in Denmark (PDF). section 6.8, p. 22, Danish Energy Agency. November 2009. ISBN 978-87-7844-821-7
- Meyers, Johan; Meneveau, Charles (1 March 2012). "Optimal turbine spacing in fully developed wind farm boundary layers". Wind Energy. 15 (2): 305–317. doi:10.1002/we.469
- "Portugal atingiu valor recorde do século na produção de eletricidade renovável e de emissões de CO2 evitadas". QUERCUS. Retrieved 14 August 2016
- Navid Goudarzi (June 2013). "A Review on the Development of the Wind Turbine Generators across the World". International Journal of Dynamics and Control. Springer. 1 (2): 192–202. doi:10.1007/s40435-013-0016-y
- Laurence Turner, Roy Gregory (2009). Windmills of Yorkshire. Catrine, East Ayrshire: Stenlake Publishing. p. 2. ISBN 9781840334753
- Zavadil, R.; Miller, N.; Ellis, A.; Muljadi, E. (2005). "Making connections". Power and Energy Magazine, IEEE. 3 (6): 26–37. doi:10.1109/MPAE.2005.1524618
- "LBNL/NREL Analysis Predicts Record Low LCOE for Wind Energy in 2012–2013". US Department of Energy Wind Program Newsletter. Archived from the original on 5 March 2012. Retrieved 10 March 2012
- Young, Kathryn (3 August 2007). "Canada wind farms blow away turbine tourists". Edmonton Journal. Retrieved 6 September 2008
- Tucker, M.J.; Pitt, E.G. (2001). "2". In Bhattacharyya, R.; McCormick, M.E. Waves in ocean engineering (1st ed.). Oxford: Elsevier. pp. 35–36. ISBN 0080435661.
- Guezuraga, Begoña; Zauner, Rudolf; Pölz, Werner (2012). "Life cycle assessment of two different 2 MW class wind turbines". Renewable Energy. 37: 37. doi:10.1016/j.renene.2011.05.008
- Hamer, Mick (21 January 2006). "The Rooftop Power Revolution". New Scientist. Reed Business Information Ltd. (2535). Retrieved 11 April 2012
- Graham, Karen." First wave-produced power in U.S. goes online in Hawaii" Digital Journal. 19 September 2016. Web Accessed 22 September 2016
- Bent Sorensen (November 1995), "History of, and Recent Progress in, Wind-Energy Utilization", Annual Review of Energy and the Environment, 20 (1): 387–424, doi:10.1146/annurev.eg.20.110195.002131
- Eilperin, Juliet; Steven Mufson (16 April 2009). "Renewable Energy's Environmental Paradox". The Washington Post. Retrieved 17 April 2009

A Brief Introduction to Solar Energy

Solar energy is generated by technologies like solar thermal energy, solar heating, solar architecture and photovoltaics. Solar energy is an appealing source of electricity as it is renewable and economical. This chapter has been carefully written to provide an easy understanding of the varied facets of solar energy.

Solar Energy

Solar energy is radiant light and heat from the Sun that is harnessed using a range of ever-evolving technologies such as solar heating, photovoltaics, solar thermal energy, solar architecture, molten salt power plants and artificial photosynthesis.

The source of Earth's solar power: the Sun

It is an important source of renewable energy and its technologies are broadly characterized as either passive solar or active solar depending on how they capture and distribute solar energy or convert it into solar power. Active solar techniques include the use of photovoltaic systems, concentrated solar power and solar water heating to harness the energy. Passive solar techniques include orienting a building to the Sun, selecting materials with favorable thermal mass or light-dispersing properties, and designing spaces that naturally circulate air.

The large magnitude of solar energy available makes it a highly appealing source of electricity. The United Nations Development Programme in its 2000 World Energy Assessment found that the annual potential of solar energy was 1,575–49,837 exajoules (EJ). This is several times larger than the total world energy consumption, which was 559.8 EJ in 2012.

In 2011, the International Energy Agency said that "the development of affordable, inexhaustible and clean solar energy technologies will have huge longer-term benefits. It will increase countries' energy security through reliance on an indigenous, inexhaustible and mostly import-independent resource, enhance sustainability, reduce pollution, lower the costs of mitigating global warming, and keep fossil fuel prices lower than otherwise. These advantages are global. Hence the additional costs of the incentives for early deployment should be considered learning investments; they must be wisely spent and need to be widely shared".

Potential

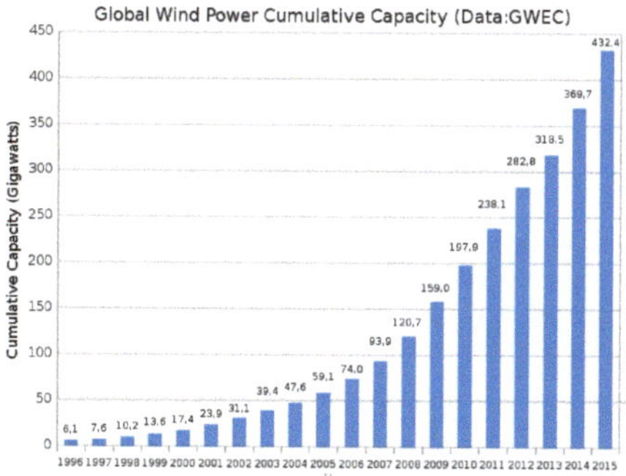

About half the incoming solar energy reaches the Earth's surface.

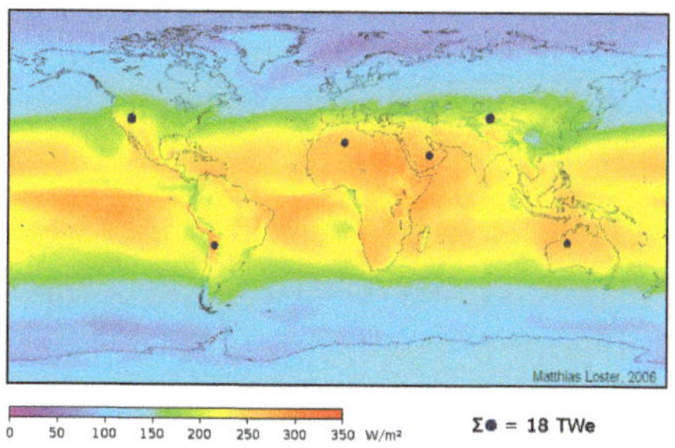

Average insolation. The theoretical area of the small black dots is sufficient to supply the world's total energy needs of 18 TW with solar power.

The Earth receives 174 petawatts (TW) of incoming solar radiation (insolation) at the upper atmosphere. Approximately 30% is reflected back to space while the rest is absorbed by clouds, oceans and land masses. The spectrum of solar light at the Earth's surface is mostly spread across the visible and near-infrared ranges with a small part in the near-ultraviolet. Most of the world's population live in areas with insolation levels of 150-300 watts/m^2, or 3.5-7.0 kWh/m^2 per day.

Solar radiation is absorbed by the Earth's land surface, oceans – which cover about 71% of the globe – and atmosphere. Warm air containing evaporated water from the oceans rises, causing atmospheric circulation or convection. When the air reaches a high altitude, where the temperature is low, water vapor condenses into clouds, which rain onto the Earth's surface, completing the water cycle. The latent heat of water condensation amplifies convection, producing atmospheric phenomena such as wind, cyclones and anti-cyclones. Sunlight absorbed by the oceans and land masses keeps the surface at an average temperature of 14 °C. By photosynthesis, green plants convert solar energy into chemically stored energy, which produces food, wood and the biomass from which fossil fuels are derived.

The total solar energy absorbed by Earth's atmosphere, oceans and land masses is approximately 3,850,000 exajoules (EJ) per year. In 2002, this was more energy in one hour than the world used in one year. Photosynthesis captures approximately 3,000 EJ per year in biomass. The amount of solar energy reaching the surface of the planet is so vast that in one year it is about twice as much as will ever be obtained from all of the Earth's non-renewable resources of coal, oil, natural gas, and mined uranium combined,

Yearly solar fluxes & human consumption[1]		
Solar	3,850,000	
Wind	2,250	
Biomass potential	~200	
Primary energy use[2]	539	
Electricity[2]	~67	
[1] Energy given in Exajoule (EJ) = 10^{18} J = 278 TWh [2] Consumption as of year 2010		

The potential solar energy that could be used by humans differs from the amount of solar energy present near the surface of the planet because factors such as geography, time variation, cloud cover, and the land available to humans limit the amount of solar energy that we can acquire.

Geography affects solar energy potential because areas that are closer to the equator have a greater amount of solar radiation. However, the use of photovoltaics that can follow the position of the sun can significantly increase the solar energy potential in areas that are farther from the equator. Time variation effects the potential of solar energy because during the nighttime there is little solar radiation on the surface of the Earth for solar panels to absorb. This limits the amount of energy that solar panels can absorb in one day. Cloud cover can affect the potential of solar panels because clouds block incoming light from the sun and reduce the light available for solar cells.

In addition, land availability has a large effect on the available solar energy because solar panels can only be set up on land that is otherwise unused and suitable for solar panels. Roofs have been found to be a suitable place for solar cells, as many people have discovered that they can collect energy directly from their homes this way. Other areas that are suitable for solar cells are lands that are not being used for businesses where solar plants can be established.

Solar technologies are characterized as either passive or active depending on the way they capture, convert and distribute sunlight and enable solar energy to be harnessed at different levels around the world, mostly depending on distance from the equator. Although solar energy refers primarily to the use of solar radiation for practical ends, all renewable energies, other than Geothermal power and Tidal power, derive their energy either directly or indirectly from the Sun.

Active solar techniques use photovoltaics, concentrated solar power, solar thermal collectors, pumps, and fans to convert sunlight into useful outputs. Passive solar techniques include selecting materials with favorable thermal properties, designing spaces that naturally circulate air, and referencing the position of a building to the Sun. Active solar technologies increase the supply of energy and are considered supply side technologies, while passive solar technologies reduce the need for alternate resources and are generally considered demand side technologies.

In 2000, the United Nations Development Programme, UN Department of Economic and Social Affairs, and World Energy Council published an estimate of the potential solar energy that could be used by humans each year that took into account factors such as insolation, cloud cover, and the land that is usable by humans. The estimate found that solar energy has a global potential of 1,575–49,837 EJ per year.

Annual solar energy potential by region (Exajoules)											
Region	North America	Latin America and Caribbean	Western Europe	Central and Eastern Europe	Former Soviet Union	Middle East and North Africa	Sub-Saharan Africa	Pacific Asia	South Asia	Centrally planned Asia	Pacific OECD
Minimum	181.1	112.6	25.1	4.5	199.3	412.4	371.9	41.0	38.8	115.5	72.6
Maximum	7,410	3,385	914	154	8,655	11,060	9,528	994	1,339	4,135	2,263

Quantitative relation of global solar potential vs. the world's primary energy consumption:

- Ratio of potential vs. current consumption (402 EJ) as of year: 3.9 (minimum) to 124 (maximum)
- Ratio of potential vs. projected consumption by 2050 (590–1,050 EJ): 1.5–2.7 (minimum) to 47–84 (maximum)
- Ratio of potential vs. projected consumption by 2100 (880–1,900 EJ): 0.8–1.8 (minimum) to 26–57 (maximum)

Thermal Energy

Solar thermal technologies can be used for water heating, space heating, space cooling and process heat generation.

Early Commercial Adaptation

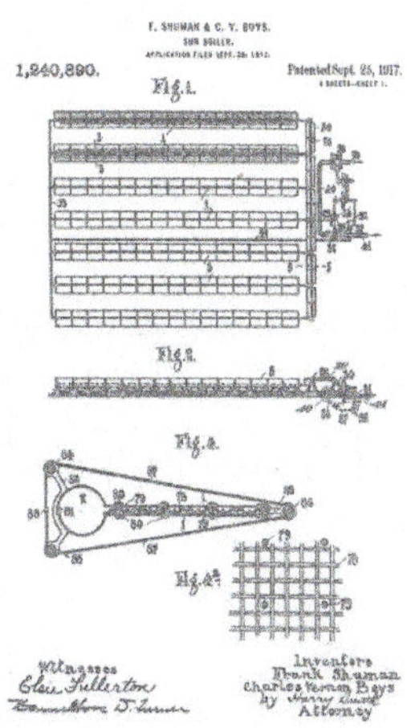

1917 Patent drawing of Shuman's solar collector

In 1897, Frank Shuman, a U.S. inventor, engineer and solar energy pioneer built a small demonstration solar engine that worked by reflecting solar energy onto square boxes filled with ether, which has a lower boiling point than water, and were fitted internally with black pipes which in turn powered a steam engine. In 1908 Shuman formed the Sun Power Company with the intent of building larger solar power plants. He, along with his technical advisor A.S.E. Ackermann and

British physicist Sir Charles Vernon Boys, developed an improved system using mirrors to reflect solar energy upon collector boxes, increasing heating capacity to the extent that water could now be used instead of ether. Shuman then constructed a full-scale steam engine powered by low-pressure water, enabling him to patent the entire solar engine system by 1912.

Shuman built the world's first solar thermal power station in Maadi, Egypt, between 1912 and 1913. His plant used parabolic troughs to power a 45–52 kilowatts (60–70 hp) engine that pumped more than 22,000 litres (4,800 imp gal; 5,800 US gal) of water per minute from the Nile River to adjacent cotton fields. Although the outbreak of World War I and the discovery of cheap oil in the 1930s discouraged the advancement of solar energy, Shuman's vision and basic design were resurrected in the 1970s with a new wave of interest in solar thermal energy. In 1916 Shuman was quoted in the media advocating solar energy's utilization, saying.

We have proved the commercial profit of sun power in the tropics and have more particularly proved that after our stores of oil and coal are exhausted the human race can receive unlimited power from the rays of the sun.

—Frank Shuman, New York Times, 2 July 1916

Water Heating

Solar water heaters facing the Sun to maximize gain

Solar hot water systems use sunlight to heat water. In low geographical latitudes (below 40 degrees) from 60 to 70% of the domestic hot water use with temperatures up to 60 °C can be provided by solar heating systems. The most common types of solar water heaters are evacuated tube collectors (44%) and glazed flat plate collectors (34%) generally used for domestic hot water; and unglazed plastic collectors (21%) used mainly to heat swimming pools.

As of 2007, the total installed capacity of solar hot water systems was approximately 154 thermal gigawatt (GW_{th}). China is the world leader in their deployment with 70 GW_{th} installed as of 2006 and a long-term goal of 210 GW_{th} by 2020. Israel and Cyprus are the per capita leaders in the use of solar hot water systems with over 90% of homes using them. In the United States, Canada, and Australia, heating swimming pools is the dominant application of solar hot water with an installed capacity of 18 GW_{th} as of 2005.

Heating, Cooling and Ventilation

In the United States, heating, ventilation and air conditioning (HVAC) systems account for 30% (4.65 EJ/yr) of the energy used in commercial buildings and nearly 50% (10.1 EJ/yr) of the energy used in residential buildings. Solar heating, cooling and ventilation technologies can be used to offset a portion of this energy.

MIT's Solar House #1, built in 1939 in the U.S., used seasonal thermal energy storage for year-round heating.

Thermal mass is any material that can be used to store heat—heat from the Sun in the case of solar energy. Common thermal mass materials include stone, cement and water. Historically they have been used in arid climates or warm temperate regions to keep buildings cool by absorbing solar energy during the day and radiating stored heat to the cooler atmosphere at night. However, they can be used in cold temperate areas to maintain warmth as well. The size and placement of thermal mass depend on several factors such as climate, daylighting and shading conditions. When properly incorporated, thermal mass maintains space temperatures in a comfortable range and reduces the need for auxiliary heating and cooling equipment.

A solar chimney (or thermal chimney, in this context) is a passive solar ventilation system composed of a vertical shaft connecting the interior and exterior of a building. As the chimney warms, the air inside is heated causing an updraft that pulls air through the building. Performance can be improved by using glazing and thermal mass materials in a way that mimics greenhouses.

Deciduous trees and plants have been promoted as a means of controlling solar heating and cooling. When planted on the southern side of a building in the northern hemisphere or the northern side in the southern hemisphere, their leaves provide shade during the summer, while the bare limbs allow light to pass during the winter. Since bare, leafless trees shade 1/3 to 1/2 of incident solar radiation, there is a balance between the benefits of summer shading and the corresponding loss of winter heating. In climates with significant heating loads, deciduous trees should not be planted on the Equator-facing side of a building because they will interfere with winter solar availability. They can, however, be used on the east and west sides to provide a degree of summer shading without appreciably affecting winter solar gain.

Cooking

Parabolic dish produces steam for cooking, in Auroville, India

Solar cookers use sunlight for cooking, drying and pasteurization. They can be grouped into three broad categories: box cookers, panel cookers and reflector cookers. The simplest solar cooker is the box cooker first built by Horace de Saussure in 1767. A basic box cooker consists of an insulated container with a transparent lid. It can be used effectively with partially overcast skies and will typically reach temperatures of 90–150 °C (194–302 °F). Panel cookers use a reflective panel to direct sunlight onto an insulated container and reach temperatures comparable to box cookers. Reflector cookers use various concentrating geometries (dish, trough, Fresnel mirrors) to focus light on a cooking container. These cookers reach temperatures of 315 °C (599 °F) and above but require direct light to function properly and must be repositioned to track the Sun.

Process Heat

Solar concentrating technologies such as parabolic dish, trough and Scheffler reflectors can provide process heat for commercial and industrial applications. The first commercial system was the Solar Total Energy Project (STEP) in Shenandoah, Georgia, USA where a field of 114 parabolic dishes provided 50% of the process heating, air conditioning and electrical requirements for a clothing factory. This grid-connected cogeneration system provided 400 kW of electricity plus thermal energy in the form of 401 kW steam and 468 kW chilled water, and had a one-hour peak load thermal storage. Evaporation ponds are shallow pools that concentrate dissolved solids through evaporation. The use of evaporation ponds to obtain salt from seawater is one of the oldest applications of solar energy. Modern uses include concentrating brine solutions used in leach mining and removing dissolved solids from waste streams. Clothes lines, clotheshorses, and clothes racks dry clothes through evaporation by wind and sunlight without consuming electricity or gas. In some states of the United States legislation protects the "right to dry" clothes. Unglazed transpired collectors (UTC) are perforated sun-facing walls used for preheating ventilation air. UTCs can raise the incoming air temperature up to 22 °C (40 °F) and deliver outlet temperatures of 45–60 °C (113–140 °F). The short payback period of transpired collectors (3 to 12 years) makes them a more cost-effective alternative than glazed collection systems. As of 2003, over 80 systems with a combined collector area of 35,000 square metres (380,000 sq ft) had been installed worldwide, including an 860 m² (9,300 sq ft) collector in Costa Rica used for drying coffee beans and a 1,300 m² (14,000 sq ft) collector in Coimbatore, India, used for drying marigolds.

Water Treatment

Solar water disinfection in Indonesia

Solar distillation can be used to make saline or brackish water potable. The first recorded instance of this was by 16th-century Arab alchemists. A large-scale solar distillation project was first constructed in 1872 in the Chilean mining town of Las Salinas. The plant, which had solar collection area of 4,700 m² (51,000 sq ft), could produce up to 22,700 L (5,000 imp gal; 6,000 US gal) per day and operate for 40 years. Individual still designs include single-slope, double-slope (or greenhouse type), vertical, conical, inverted absorber, multi-wick, and multiple effect. These stills can operate in passive, active, or hybrid modes. Double-slope stills are the most economical for decentralized domestic purposes, while active multiple effect units are more suitable for large-scale applications.

Solar water disinfection (SODIS) involves exposing water-filled plastic polyethylene terephthalate (PET) bottles to sunlight for several hours. Exposure times vary depending on weather and climate from a minimum of six hours to two days during fully overcast conditions. It is recommended by the World Health Organization as a viable method for household water treatment and safe storage. Over two million people in developing countries use this method for their daily drinking water.

Solar energy may be used in a water stabilization pond to treat waste water without chemicals or electricity. A further environmental advantage is that algae grow in such ponds and consume carbon dioxide in photosynthesis, although algae may produce toxic chemicals that make the water unusable.

Molten Salt Technology

Molten salt can be employed as a thermal energy storage method to retain thermal energy collected by a solar tower or solar trough of a concentrated solar power plant, so that it can be used to generate electricity in bad weather or at night. It was demonstrated in the Solar Two project from 1995–1999. The system is predicted to have an annual efficiency of 99%, a reference to the energy retained by storing heat before turning it into electricity, versus converting heat directly into electricity. The molten salt mixtures vary. The most extended mixture contains sodium nitrate, potassium nitrate and calcium nitrate. It is non-flammable and nontoxic, and has already been used in the chemical and metals industries as a heat-transport fluid, so experience with such systems exists in non-solar applications.

The salt melts at 131 °C (268 °F). It is kept liquid at 288 °C (550 °F) in an insulated "cold" storage tank. The liquid salt is pumped through panels in a solar collector where the focused sun heats it to 566 °C (1,051 °F). It is then sent to a hot storage tank. This is so well insulated that the thermal energy can be usefully stored for up to a week.

When electricity is needed, the hot salt is pumped to a conventional steam-generator to produce superheated steam for a turbine/generator as used in any conventional coal, oil, or nuclear power plant. A 100-megawatt turbine would need a tank about 9.1 metres (30 ft) tall and 24 metres (79 ft) in diameter to drive it for four hours by this design.

Several parabolic trough power plants in Spain and solar power tower developer SolarReserve use this thermal energy storage concept. The Solana Generating Station in the U.S. has six hours of storage by molten salt. The María Elena plant is a 400 MW thermo-solar complex in the northern Chilean region of Antofagasta employing molten salt technology.

Electricity Production

Some of the world's largest solar power stations: Ivanpah (CSP) and Topaz (PV)

Solar power is the conversion of sunlight into electricity, either directly using photovoltaics (PV), or indirectly using concentrated solar power (CSP). CSP systems use lenses or mirrors and tracking systems to focus a large area of sunlight into a small beam. PV converts light into electric current using the photoelectric effect.

Solar power is anticipated to become the world's largest source of electricity by 2050, with solar photovoltaics and concentrated solar power contributing 16 and 11 percent to the global overall consumption, respectively.

Commercial CSP plants were first developed in the 1980s. Since 1985 the eventually 354 MW SEGS CSP installation, in the Mojave Desert of California, is the largest solar power plant in the world. Other large CSP plants include the 150 MW Solnova Solar Power Station and the 100 MW Andasol solar power station, both in Spain. The 250 MW Agua Caliente Solar Project, in the United

States, and the 221 MW Charanka Solar Park in India, are the world's largest photovoltaic plants. Solar projects exceeding 1 GW are being developed, but most of the deployed photovoltaics are in small rooftop arrays of less than 5 kW, which are connected to the grid using net metering and/or a feed-in tariff. In 2013 solar generated less than 1% of the world's total grid electricity.

Photovoltaics

In the last two decades, photovoltaics (PV), also known as solar PV, has evolved from a pure niche market of small scale applications towards becoming a mainstream electricity source. A solar cell is a device that converts light directly into electricity using the photoelectric effect. The first solar cell was constructed by Charles Fritts in the 1880s. In 1931 a German engineer, Dr Bruno Lange, developed a photo cell using silver selenide in place of copper oxide. Although the prototype selenium cells converted less than 1% of incident light into electricity, both Ernst Werner von Siemens and James Clerk Maxwell recognized the importance of this discovery. Following the work of Russell Ohl in the 1940s, researchers Gerald Pearson, Calvin Fuller and Daryl Chapin created the crystalline silicon solar cell in 1954. These early solar cells cost 286 USD/watt and reached efficiencies of 4.5–6%. By 2012 available efficiencies exceeded 20%, and the maximum efficiency of research photovoltaics was in excess of 40%.

Concentrated Solar Power

Concentrating Solar Power (CSP) systems use lenses or mirrors and tracking systems to focus a large area of sunlight into a small beam. The concentrated heat is then used as a heat source for a conventional power plant. A wide range of concentrating technologies exists; the most developed are the parabolic trough, the concentrating linear fresnel reflector, the Stirling dish and the solar power tower. Various techniques are used to track the Sun and focus light. In all of these systems a working fluid is heated by the concentrated sunlight, and is then used for power generation or energy storage.

Architecture and Urban Planning

Darmstadt University of Technology, Germany, won the 2007 Solar Decathlon in Washington, D.C. with this passive house designed for humid and hot subtropical climate.

Sunlight has influenced building design since the beginning of architectural history. Advanced solar architecture and urban planning methods were first employed by the Greeks and Chinese, who oriented their buildings toward the south to provide light and warmth.

The common features of passive solar architecture are orientation relative to the Sun, compact proportion (a low surface area to volume ratio), selective shading (overhangs) and thermal mass. When these features are tailored to the local climate and environment they can produce well-lit spaces that stay in a comfortable temperature range. Socrates' Megaron House is a classic example of passive solar design. The most recent approaches to solar design use computer modeling tying together solar lighting, heating and ventilation systems in an integrated solar design package. Active solar equipment such as pumps, fans and switchable windows can complement passive design and improve system performance.

Urban heat islands (UHI) are metropolitan areas with higher temperatures than that of the surrounding environment. The higher temperatures result from increased absorption of solar energy by urban materials such as asphalt and concrete, which have lower albedos and higher heat capacities than those in the natural environment. A straightforward method of counteracting the UHI effect is to paint buildings and roads white, and to plant trees in the area. Using these methods, a hypothetical "cool communities" program in Los Angeles has projected that urban temperatures could be reduced by approximately 3 °C at an estimated cost of US$1 billion, giving estimated total annual benefits of US$530 million from reduced air-conditioning costs and healthcare savings.

Agriculture and Horticulture

Greenhouses like these in the Westland municipality of the Netherlands grow vegetables, fruits and flowers.

Agriculture and horticulture seek to optimize the capture of solar energy in order to optimize the productivity of plants. Techniques such as timed planting cycles, tailored row orientation, staggered heights between rows and the mixing of plant varieties can improve crop yields. While sunlight is generally considered a plentiful resource, the exceptions highlight the importance of solar energy to agriculture. During the short growing seasons of the Little Ice Age, French and English farmers employed fruit walls to maximize the collection of solar energy. These walls acted as thermal masses and accelerated ripening by keeping plants warm. Early fruit walls were built perpendicular to the ground and facing south, but over time, sloping walls were developed to make better use of sunlight. In 1699, Nicolas Fatio de Duillier even suggested using a tracking mechanism which could pivot to follow the Sun. Applications of solar energy in agriculture aside from growing crops include pumping water, drying crops, brooding chicks and drying chicken manure. More recently the technology has been embraced by vintners, who use the energy generated by solar panels to power grape presses.

Greenhouses convert solar light to heat, enabling year-round production and the growth (in enclosed environments) of specialty crops and other plants not naturally suited to the local climate. Primitive greenhouses were first used during Roman times to produce cucumbers year-round for the Roman emperor Tiberius. The first modern greenhouses were built in Europe in the 16th century to keep exotic plants brought back from explorations abroad. Greenhouses remain an important part of horticulture today, and plastic transparent materials have also been used to similar effect in polytunnels and row covers.

Transport

Winner of the 2013 World Solar Challenge in Australia

Solar electric aircraft circumnavigating the globe in 2015

Development of a solar-powered car has been an engineering goal since the 1980s. The World Solar Challenge is a biannual solar-powered car race, where teams from universities and enterprises compete over 3,021 kilometres (1,877 mi) across central Australia from Darwin to Adelaide. In 1987, when it was founded, the winner's average speed was 67 kilometres per hour (42 mph) and by 2007 the winner's average speed had improved to 90.87 kilometres per hour (56.46 mph). The North American Solar Challenge and the planned South African Solar Challenge are comparable competitions that reflect an international interest in the engineering and development of solar powered vehicles.

Some vehicles use solar panels for auxiliary power, such as for air conditioning, to keep the interior cool, thus reducing fuel consumption.

In 1975, the first practical solar boat was constructed in England. By 1995, passenger boats incorporating PV panels began appearing and are now used extensively. In 1996, Kenichi Horie made the first solar-powered crossing of the Pacific Ocean, and the *Sun21* catamaran made the first solar-powered crossing of the Atlantic Ocean in the winter of 2006–2007. There were plans to circumnavigate the globe in 2010.

In 1974, the unmanned AstroFlight Sunrise airplane made the first solar flight. On 29 April 1979, the *Solar Riser* made the first flight in a solar-powered, fully controlled, man-carrying flying machine, reaching an altitude of 40 ft (12 m). In 1980, the *Gossamer Penguin* made the first piloted flights powered solely by photovoltaics. This was quickly followed by the *Solar Challenger* which crossed the English Channel in July 1981. In 1990 Eric Scott Raymond in 21 hops flew from California to North Carolina using solar power. Developments then turned back to unmanned aerial vehicles (UAV) with the *Pathfinder* (1997) and subsequent designs, culminating in the *Helios* which set the altitude record for a non-rocket-propelled aircraft at 29,524 metres (96,864 ft) in 2001. The *Zephyr*, developed by BAE Systems, is the latest in a line of record-breaking solar aircraft, making a 54-hour flight in 2007, and month-long flights were envisioned by 2010. As of 2016, Solar Impulse, an electric aircraft, is currently circumnavigating the globe. It is a single-seat plane powered by solar cells and capable of taking off under its own power. The design allows the aircraft to remain airborne for several days.

A solar balloon is a black balloon that is filled with ordinary air. As sunlight shines on the balloon, the air inside is heated and expands causing an upward buoyancy force, much like an artificially heated hot air balloon. Some solar balloons are large enough for human flight, but usage is generally limited to the toy market as the surface-area to payload-weight ratio is relatively high.

Fuel Production

Concentrated solar panels are getting a power boost. Pacific Northwest National Laboratory (PNNL) will be testing a new concentrated solar power system -- one that can help natural gas power plants reduce their fuel usage by up to 20 percent.

Solar chemical processes use solar energy to drive chemical reactions. These processes offset energy that would otherwise come from a fossil fuel source and can also convert solar energy into storable and transportable fuels. Solar induced chemical reactions can be divided into thermochemical or photochemical. A variety of fuels can be produced by artificial photosynthesis. The multielectron catalytic chemistry involved in making carbon-based fuels (such as methanol) from reduction of carbon dioxide is challenging; a feasible alternative is hydrogen production from protons, though use of water as the source of electrons (as plants do) requires mastering the multielectron oxidation of two water molecules to molecular oxygen. Some have envisaged working solar fuel plants in coastal metropolitan areas by 2050 – the splitting of sea water providing hydrogen to be run through adjacent fuel-cell electric power plants and the pure water by-product going directly into the municipal water system. Another vision involves all human structures covering the earth's surface (i.e., roads, vehicles and buildings) doing photosynthesis more efficiently than plants.

Hydrogen production technologies have been a significant area of solar chemical research since the 1970s. Aside from electrolysis driven by photovoltaic or photochemical cells, several thermochemical processes have also been explored. One such route uses concentrators to split water into oxygen and hydrogen at high temperatures (2,300–2,600 °C or 4,200–4,700 °F). Another approach uses the heat from solar concentrators to drive the steam reformation of natural gas thereby increasing the overall hydrogen yield compared to conventional reforming methods. Thermochemical cycles characterized by the decomposition and regeneration of reactants present another avenue for hydrogen production. The Solzinc process under development at the Weizmann Institute of Science uses a 1 MW solar furnace to decompose zinc oxide (ZnO) at temperatures above 1,200 °C (2,200 °F). This initial reaction produces pure zinc, which can subsequently be reacted with water to produce hydrogen.

Energy Storage Methods

Thermal energy storage. The Andasol CSP plant uses tanks of molten salt to store solar energy.

Thermal mass systems can store solar energy in the form of heat at domestically useful temperatures for daily or interseasonal durations. Thermal storage systems generally use readily available materials with high specific heat capacities such as water, earth and stone. Well-designed systems can lower peak demand, shift time-of-use to off-peak hours and reduce overall heating and cooling requirements.

Phase change materials such as paraffin wax and Glauber's salt are another thermal storage medium. These materials are inexpensive, readily available, and can deliver domestically useful temperatures (approximately 64 °C or 147 °F). The "Dover House" (in Dover, Massachusetts) was the first to use a Glauber's salt heating system, in 1948. Solar energy can also be stored at high temperatures using molten salts. Salts are an effective storage medium because they are low-cost, have a high specific heat capacity and can deliver heat at temperatures compatible with conventional power systems. The Solar Two project used this method of energy storage, allowing it to store 1.44 terajoules (400,000 kWh) in its 68 m^3 storage tank with an annual storage efficiency of about 99%.

Off-grid PV systems have traditionally used rechargeable batteries to store excess electricity. With grid-tied systems, excess electricity can be sent to the transmission grid, while standard grid electricity can be used to meet shortfalls. Net metering programs give household systems a credit for any electricity they deliver to the grid. This is handled by 'rolling back' the meter whenever the home produces more electricity than it consumes. If the net electricity use is below zero, the utility then rolls over the kilowatt hour credit to the next month. Other approaches involve the use of two meters, to measure electricity consumed vs. electricity produced. This is less common due to the increased installation cost of the second meter. Most standard meters accurately measure in both directions, making a second meter unnecessary.

Pumped-storage hydroelectricity stores energy in the form of water pumped when energy is available from a lower elevation reservoir to a higher elevation one. The energy is recovered when demand is high by releasing the water, with the pump becoming a hydroelectric power generator.

Development, Deployment and Economics

Participants in a workshop on sustainable development inspect solar panels at Monterrey Institute of Technology and Higher Education, Mexico City on top of a building on campus.

Beginning with the surge in coal use which accompanied the Industrial Revolution, energy consumption has steadily transitioned from wood and biomass to fossil fuels. The early development of solar technologies starting in the 1860s was driven by an expectation that coal would soon become scarce. However, development of solar technologies stagnated in the early 20th century in the face of the increasing availability, economy, and utility of coal and petroleum.

The 1973 oil embargo and 1979 energy crisis caused a reorganization of energy policies around the world and brought renewed attention to developing solar technologies. Deployment strategies focused on incentive programs such as the Federal Photovoltaic Utilization Program in the U.S. and the Sunshine Program in Japan. Other efforts included the formation of research facilities in the U.S. (SERI, now NREL), Japan (NEDO), and Germany (Fraunhofer Institute for Solar Energy Systems ISE).

Commercial solar water heaters began appearing in the United States in the 1890s. These systems saw increasing use until the 1920s but were gradually replaced by cheaper and more reliable heating fuels. As with photovoltaics, solar water heating attracted renewed attention as a result of the oil crises in the 1970s but interest subsided in the 1980s due to falling petroleum prices. Development in the solar water heating sector progressed steadily throughout the 1990s and annual growth rates have averaged 20% since 1999. Although generally underestimated, solar water heating and cooling is by far the most widely deployed solar technology with an estimated capacity of 154 GW as of 2007.

The International Energy Agency has said that solar energy can make considerable contributions to solving some of the most urgent problems the world now faces:

The development of affordable, inexhaustible and clean solar energy technologies will have huge longer-term benefits. It will increase countries' energy security through reliance on an indigenous, inexhaustible and mostly import-independent resource, enhance sustainability, reduce pollution, lower the costs of mitigating climate change, and keep fossil fuel prices lower than otherwise. These advantages are global. Hence the additional costs of the incentives for early deployment should be considered learning investments; they must be wisely spent and need to be widely shared.

In 2011, a report by the International Energy Agency found that solar energy technologies such as photovoltaics, solar hot water and concentrated solar power could provide a third of the world's energy by 2060 if politicians commit to limiting climate change. The energy from the sun could play a key role in de-carbonizing the global economy alongside improvements in energy efficiency and imposing costs on greenhouse gas emitters. "The strength of solar is the incredible variety and flexibility of applications, from small scale to big scale".

We have proved ... that after our stores of oil and coal are exhausted the human race can receive unlimited power from the rays of the sun.

—Frank Shuman, New York Times, 2 July 1916

ISO Standards

The International Organization for Standardization has established several standards relating to solar energy equipment. For example, ISO 9050 relates to glass in building while ISO 10217 relates to the materials used in solar water heaters.

Solar Thermal Energy

Solar thermal energy (STE) is a form of energy and a technology for harnessing solar energy to generate thermal energy or electrical energy for use in industry, and in the residential and commercial sectors.

Roof-mounted close-coupled thermosiphon solar water heater.

The first three units of Solnova in the foreground, with the two towers of the PS10 and PS20 solar power stations in the background.

Overview

Solar thermal collectors are classified by the United States Energy Information Administration as low-, medium-, or high-temperature collectors. Low-temperature collectors are flat plates generally used to heat swimming pools. Medium-temperature collectors are also usually flat plates but are used for heating water or air for residential and commercial use. High-temperature collectors concentrate sunlight using mirrors or lenses and are generally used for fulfilling heat requirements up to 300 deg C / 20 bar pressure in industries, and for electric power production. Two categories include Concentrated Solar Thermal (CST) for fulfilling heat requirements in industries, and Concentrated Solar Power (CSP) when the heat collected is used for power generation. CST and CSP are not replaceable in terms of application. The largest facilities are located in the American Mojave Desert of California and Nevada. These plants employ a variety of different technologies. The largest examples include, Ivanpah Solar Power Facility (377 MW), Solar Energy Generating Systems installation (354 MW), and Crescent Dunes (110 MW). Spain is the other major developer of solar thermal power plant. The largest examples include, Solnova Solar Power Station (150 MW), the Andasol solar power station (150 MW), and Extresol Solar Power Station (100 MW).

History

Augustin Mouchot demonstrated a solar collector with a cooling engine making ice cream at the 1878 Universal Exhibition in Paris. The first installation of solar thermal energy equipment occurred in the Sahara approximately in 1910 by Frank Shuman when a steam engine was run on steam produced by sunlight. Because liquid fuel engines were developed and found more convenient, the Sahara project was abandoned, only to be revisited several decades later.

Low-temperature Solar Heating and Cooling Systems

Systems for utilizing low-temperature solar thermal energy include means for heat collection; usually heat storage, either short-term or interseasonal; and distribution within a structure or a district heating network. In some cases more than one of these functions is inherent to a single feature of the system (e.g. some kinds of solar collectors also store heat). Some systems are passive, others are active (requiring other external energy to function).

Heating is the most obvious application, but solar cooling can be achieved for a building or district cooling network by using a heat-driven absorption or adsorption chiller (heat pump). There is a productive coincidence that the greater the driving heat from insulation, the greater the cooling output. In 1878, Auguste Mouchout pioneered solar cooling by making ice using a solar steam engine attached to a refrigeration device.

MIT's Solar House #1 built in 1939 used seasonal thermal energy storage (STES) for year-round heating.

In the United States, heating, ventilation, and air conditioning (HVAC) systems account for over 25% (4.75 EJ) of the energy used in commercial buildings and nearly half (10.1 EJ) of the energy used in residential buildings. Solar heating, cooling, and ventilation technologies can be used to offset a portion of this energy.

In Europe, since the mid-1990s about 125 large solar-thermal district heating plants have been constructed, each with over 500 m² (5400 ft²) of solar collectors. The largest are about 10,000 m², with capacities of 7 MW-thermal and solar heat costs around 4 Eurocents/kWh without subsidies. 40 of them have nominal capacities of 1 MW-thermal or more. The Solar District Heating program (SDH) has participation from 14 European Nations and the European Commission, and is working toward technical and market development, and holds annual conferences.

Low-temperature Collectors

Glazed solar collectors are designed primarily for space heating. They recirculate building air through a solar air panel where the air is heated and then directed back into the building. These solar space heating systems require at least two penetrations into the building and only perform when the air in the solar collector is warmer than the building room temperature. Most glazed collectors are used in the residential sector.

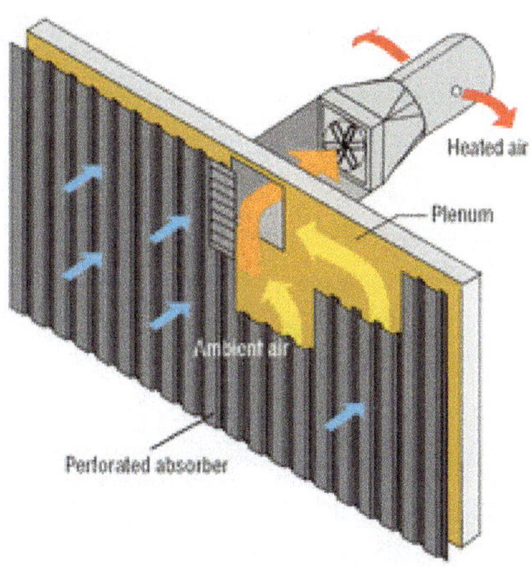

Unglazed, "transpired" air collector

Unglazed solar collectors are primarily used to pre-heat make-up ventilation air in commercial, industrial and institutional buildings with a high ventilation load. They turn building walls or sections of walls into low cost, high performance, unglazed solar collectors. Also called, "transpired solar panels" or "solar wall", they employ a painted perforated metal solar heat absorber that also serves as the exterior wall surface of the building. Heat conducts from the absorber surface to the thermal boundary layer of air 1 mm thick on the outside of the absorber and to air that passes behind the absorber. The boundary layer of air is drawn into a nearby perforation before the heat can escape by convection to the outside air. The heated air is then drawn from behind the absorber plate into the building's ventilation system.

A Trombe wall is a passive solar heating and ventilation system consisting of an air channel sandwiched between a window and a sun-facing thermal mass. During the ventilation cycle, sunlight stores heat in the thermal mass and warms the air channel causing circulation through vents at the top and bottom of the wall. During the heating cycle the Trombe wall radiates stored heat.

Solar roof ponds are unique solar heating and cooling systems developed by Harold Hay in the 1960s. A basic system consists of a roof-mounted water bladder with a movable insulating cover. This system can control heat exchange between interior and exterior environments by covering and uncovering the bladder between night and day. When heating is a concern the bladder is uncovered during the day allowing sunlight to warm the water bladder and store heat for evening use. When cooling is a concern the covered bladder draws heat from the building's interior during the day and is uncovered at night to radiate heat to the cooler atmosphere. The Skytherm house in Atascadero, California uses a prototype roof pond for heating and cooling.

Solar space heating with solar air heat collectors is more popular in the USA and Canada than heating with solar liquid collectors since most buildings already have a ventilation system for heating and cooling. The two main types of solar air panels are glazed and unglazed.

Of the 21,000,000 square feet (2,000,000 m^2) of solar thermal collectors produced in the United States in 2007, 16,000,000 square feet (1,500,000 m^2) were of the low-temperature variety. Low-temperature collectors are generally installed to heat swimming pools, although they can also be used for space heating. Collectors can use air or water as the medium to transfer the heat to their destination.

Heat Storage in Low-temperature Solar Thermal Systems

Interseasonal storage: Solar heat (or heat from other sources) can be effectively stored between opposing seasons in aquifers, underground geological strata, large specially constructed pits, and large tanks that are insulated and covered with earth.

Short-term storage. Thermal mass materials store solar energy during the day and release this energy during cooler periods. Common thermal mass materials include stone, concrete, and water. The proportion and placement of thermal mass should consider several factors such as climate, daylighting, and shading conditions. When properly incorporated, thermal mass can passively maintain comfortable temperatures while reducing energy consumption.

Solar-driven Cooling

Worldwide, by 2011 there were about 750 cooling systems with solar-driven heat pumps, and annual market growth was 40 to 70% over the prior seven years. It is a niche market because the economics are challenging, with the annual number of cooling hours a limiting factor. Respectively, the annual cooling hours are roughly 1000 in the Mediterranean, 2500 in Southeast Asia, and only 50 to 200 in Central Europe. However, system construction costs dropped about 50% between 2007 and 2011. The International Energy Agency (IEA) Solar Heating and Cooling program (IEA-SHC) task groups working on further development of the technologies involved.

Solar Heat-driven Ventilation

A solar chimney (or thermal chimney) is a passive solar ventilation system composed of a hollow thermal mass connecting the interior and exterior of a building. As the chimney warms, the air inside is heated causing an updraft that pulls air through the building. These systems have been in use since Roman times and remain common in the Middle East.

Process Heat

Solar Evaporation Ponds in the Atacama Desert.

Solar process heating systems are designed to provide large quantities of hot water or space heating for nonresidential buildings.

Evaporation ponds are shallow ponds that concentrate dissolved solids through evaporation. The use of evaporation ponds to obtain salt from sea water is one of the oldest applications of solar energy. Modern uses include concentrating brine solutions used in leach mining and removing dissolved solids from waste streams. Altogether, evaporation ponds represent one of the largest commercial applications of solar energy in use today.

Unglazed transpired collectors are perforated sun-facing walls used for preheating ventilation air. Transpired collectors can also be roof mounted for year-round use and can raise the incoming air temperature up to 22 °C and deliver outlet temperatures of 45-60 °C. The short payback period of transpired collectors (3 to 12 years) make them a more cost-effective alternative to glazed collection systems. As of 2015, over 4000 systems with a combined collector area of 500,000 m² had been installed worldwide. Representatives include an 860 m² collector in Costa Rica used for drying coffee beans and a 1300 m² collector in Coimbatore, India used for drying marigolds.

A food processing facility in Modesto, California uses parabolic troughs to produce steam used in the manufacturing process. The 5,000 m² collector area is expected to provide 15 TJ per year.

Medium-temperature Collectors

These collectors could be used to produce approximately 50% and more of the hot water needed for residential and commercial use in the United States. In the United States, a typical system costs $4000–$6000 retail ($1400 to $2200 wholesale for the materials) and 30% of the system qualifies for a federal tax credit + additional state credit exists in about half of the states. Labor for a simple open loop system in southern climates can take 3–5 hours for the installation and 4–6 hours in Northern areas. Northern system require more collector area and more complex plumbing to protect the collector from freezing. With this incentive, the payback time for a typical household is four to nine years, depending on the state. Similar subsidies exist in parts of Europe. A crew of one solar plumber and two assistants with minimal training can install a system per day. Thermosiphon installation have negligible maintenance costs (costs rise if antifreeze and mains power are used for circulation) and in the US reduces a households' operating costs by $6 per person per month. Solar water heating can reduce CO_2 emissions of a family of four by 1 ton/year (if replacing natural gas) or 3 ton/year (if replacing electricity). Medium-temperature installations can use any of several designs: common designs are pressurized glycol, drain back, batch systems and newer low pressure freeze tolerant systems using polymer pipes containing water with photovoltaic pumping. European and International standards are being reviewed to accommodate innovations in design and operation of medium temperature collectors. Operational innovations include "permanently wetted collector" operation. This innovation reduces or even eliminates the occurrence of no-flow high temperature stresses called stagnation which would otherwise reduce the life expectancy of collectors.

Solar Drying

Industrial indirect solar fruit and vegetable dryer

Solar thermal energy can be useful for drying wood for construction and wood fuels such as wood chips for combustion. Solar is also used for food products such as fruits, grains, and fish. Crop drying by solar means is environmentally friendly as well as cost effective while improving the quality. The less money it takes to make a product, the less it can be sold for, pleasing both the buyers and the sellers. Technologies in solar drying include ultra low cost pumped transpired plate air collectors based on black fabrics. Solar thermal energy is helpful in the process of drying products such as wood chips and other forms of biomass by raising the temperature while allowing air to pass through and get rid of the moisture.

Cooking

The Solar Bowl above the Solar Kitchen in Auroville, India concentrates sunlight on a movable receiver to produce steam for cooking.

Solar cookers use sunlight for cooking, drying and pasteurization. Solar cooking offsets fuel costs, reduces demand for fuel or firewood, and improves air quality by reducing or removing a source of smoke.

The simplest type of solar cooker is the box cooker first built by Horace de Saussure in 1767. A basic box cooker consists of an insulated container with a transparent lid. These cookers can be used effectively with partially overcast skies and will typically reach temperatures of 50–100 °C.

Concentrating solar cookers use reflectors to concentrate solar energy onto a cooking container. The most common reflector geometries are flat plate, disc and parabolic trough type. These designs cook faster and at higher temperatures (up to 350 °C) but require direct light to function properly.

The Solar Kitchen in Auroville, India uses a unique concentrating technology known as the solar bowl. Contrary to conventional tracking reflector/fixed receiver systems, the solar bowl uses a fixed spherical reflector with a receiver which tracks the focus of light as the Sun moves across the sky. The solar bowl's receiver reaches temperature of 150 °C that is used to produce steam that helps cook 2,000 daily meals.

Many other solar kitchens in India use another unique concentrating technology known as the Scheffler reflector. This technology was first developed by Wolfgang Scheffler in 1986. A Scheffler reflector is a parabolic dish that uses single axis tracking to follow the Sun's daily course. These reflectors have a flexible reflective surface that is able to change its curvature to adjust to seasonal variations in the incident angle of sunlight. Scheffler reflectors have the advantage of having a fixed focal point which improves the ease of cooking and are able to reach temperatures of 450-650 °C. Built in 1999 by the Brahma Kumaris, the world's largest Scheffler reflector system in Abu Road, Rajasthan India is capable of cooking up to 35,000 meals a day. By early 2008, over 2000 large cookers of the Scheffler design had been built worldwide.

Distillation

Solar stills can be used to make drinking water in areas where clean water is not common. Solar distillation is necessary in these situations to provide people with purified water. Solar energy heats up the water in the still. The water then evaporates and condenses on the bottom of the covering glass.

High-temperature Collectors

Part of the 354 MW SEGS solar complex in northern San Bernardino County, California.

The solar furnace at Odeillo in the French Pyrenees-Orientales can reach temperatures up to 3,500°C .

Where temperatures below about 95 °C are sufficient, as for space heating, flat-plate collectors of the nonconcentrating type are generally used. Because of the relatively high heat losses through the glazing, flat plate collectors will not reach temperatures much above 200 °C even when the heat transfer fluid is stagnant. Such temperatures are too low for efficient conversion to electricity.

The efficiency of heat engines increases with the temperature of the heat source. To achieve this in solar thermal energy plants, solar radiation is concentrated by mirrors or lenses to obtain higher temperatures – a technique called Concentrated Solar Power (CSP). The practical effect of high efficiencies is to reduce the plant's collector size and total land use per unit power generated, reducing the environmental impacts of a power plant as well as its expense.

As the temperature increases, different forms of conversion become practical. Up to 600 °C, steam turbines, standard technology, have an efficiency up to 41%. Above 600 °C, gas turbines can be more efficient. Higher temperatures are problematic because different materials and techniques are needed. One proposal for very high temperatures is to use liquid fluoride salts operating between 700 °C to 800 °C, using multi-stage turbine systems to achieve 50% or more thermal efficiencies. The higher operating temperatures permit the plant to use higher-temperature dry heat exchangers for its thermal exhaust, reducing the plant's water use – critical in the deserts where large solar plants are practical. High temperatures also make heat storage more efficient, because more watt-hours are stored per unit of fluid.

Commercial concentrating solar thermal power (CSP) plants were first developed in the 1980s.

The world's largest solar thermal power plants are now the 370 MW Ivanpah Solar Power Facility, commissioned in 2014, and the 354 MW SEGS CSP installation, both located in the Mojave Desert of California, where several other solar projects have been realized as well. With the exception of the Shams solar power station, built in 2013 near Abu Dhabi, the United Arab Emirates, all other 100 MW or larger CSP plants are either located in the United States or in Spain.

The principal advantage of CSP is the ability to efficiently add thermal storage, allowing the dispatching of electricity over up to a 24-hour period. Since peak electricity demand typically occurs at about 5 pm, many CSP power plants use 3 to 5 hours of thermal storage. With current technology, storage of heat is much cheaper and more efficient than storage of electricity. In this way, the CSP plant can produce electricity day and night. If the CSP site has predictable solar radiation, then the CSP plant becomes a reliable power plant. Reliability can further be improved by installing a back-up combustion system. The back-up system can use most of the CSP plant, which decreases the cost of the back-up system.

CSP facilities utilize high electrical conductivity materials, such as copper, in field power cables, grounding networks, and motors for tracking and pumping fluids, as well as in the main generator and high voltage transformers.

With reliability, unused desert, no pollution, and no fuel costs, the obstacles for large deployment for CSP are cost, aesthetics, land use and similar factors for the necessary connecting high tension lines. Although only a small percentage of the desert is necessary to meet global electricity demand, still a large area must be covered with mirrors or lenses to obtain a significant amount of energy. An important way to decrease cost is the use of a simple design.

When considering land use impacts associated with the exploration and extraction through to transportation and conversion of fossil fuels, which are used for most of our electrical power, utility-scale solar power compares as one of the most land-efficient energy resources available:

The federal government has dedicated nearly 2,000 times more acreage to oil and gas leases than to solar development. In 2010 the Bureau of Land Management approved nine large-scale solar projects, with a total generating capacity of 3,682 megawatts, representing approximately 40,000 acres. In contrast, in 2010, the Bureau of Land Management processed more than 5,200 applications gas and oil leases, and issued 1,308 leases, for a total of 3.2 million acres. Currently, 38.2 million acres of onshore public lands and an additional 36.9 million acres of offshore exploration in the Gulf of Mexico are under lease for oil and gas development, exploration and production.

System Designs

During the day the sun has different positions. For low concentration systems (and low temperatures) tracking can be avoided (or limited to a few positions per year) if nonimaging optics are used. For higher concentrations, however, if the mirrors or lenses do not move, then the focus of the mirrors or lenses changes (but also in these cases nonimaging optics provides the widest acceptance angles for a given concentration). Therefore, it seems unavoidable that there needs to be a tracking system that follows the position of the sun (for solar photovoltaic a solar tracker is only optional). The tracking system increases the cost and complexity. With this in mind, different designs can be distinguished in how they concentrate the light and track the position of the sun.

Parabolic Trough Designs

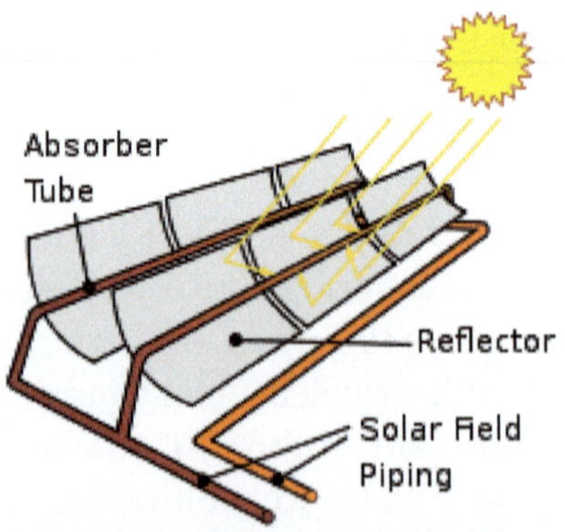

Sketch of a parabolic trough design. A change of position of the sun parallel to the receiver does not require adjustment of the mirrors.

Parabolic trough power plants use a curved, mirrored trough which reflects the direct solar radiation onto a glass tube containing a fluid (also called a receiver, absorber or collector) running the length of the trough, positioned at the focal point of the reflectors. The trough is parabolic along one axis and linear in the orthogonal axis. For change of the daily position of the sun perpendicular to the receiver, the trough tilts east to west so that the direct radiation remains focused on the receiver. However, seasonal changes in the angle of sunlight parallel to the trough does not require adjustment of the mirrors, since the light is simply concentrated elsewhere on the receiver. Thus the trough design does not require tracking on a second axis. The receiver may be enclosed in a glass vacuum chamber. The vacuum significantly reduces convective heat loss.

A fluid (also called heat transfer fluid) passes through the receiver and becomes very hot. Common fluids are synthetic oil, molten salt and pressurized steam. The fluid containing the heat is transported to a heat engine where about a third of the heat is converted to electricity.

Full-scale parabolic trough systems consist of many such troughs laid out in parallel over a large area of land. Since 1985 a solar thermal system using this principle has been in full operation in California in the United States. It is called the Solar Energy Generating Systems (SEGS) system. Other CSP designs lack this kind of long experience and therefore it can currently be said that the parabolic trough design is the most thoroughly proven CSP technology.

The SEGS is a collection of nine plants with a total capacity of 354 MW and has been the world's largest solar power plant, both thermal and non-thermal, for many years. A newer plant is Nevada Solar One plant with a capacity of 64 MW. The 150 MW Andasol solar power stations are in Spain with each site having a capacity of 50 MW. Note however, that those plants have heat storage which requires a larger field of solar collectors relative to the size of the steam turbine-generator to store heat and send heat to the steam turbine at the same time. Heat storage enables better utilization of the steam turbine. With day and some nighttime operation of the steam-turbine Andasol 1 at 50 MW peak capacity produces more energy than Nevada Solar One at 64 MW peak capacity, due to the former plant's thermal energy storage system and larger solar field. The 280MW Solana Generating Station came online in Arizona in 2013 with 6 hours of power storage. Hassi R'Mel integrated solar combined cycle power station in Algeria and Martin Next Generation Solar Energy Center both use parabolic troughs in a combined cycle with natural gas.

Enclosed Trough

Inside an enclosed trough system

The enclosed trough architecture encapsulates the solar thermal system within a greenhouse-like glasshouse. The glasshouse creates a protected environment to withstand the elements that can negatively impact reliability and efficiency of the solar thermal system.

Lightweight curved solar-reflecting mirrors are suspended within the glasshouse structure. A single-axis tracking system positions the mirrors to track the sun and focus its light onto a network of stationary steel pipes, also suspended from the glasshouse structure. Steam is generated directly, using oil field-quality water, as water flows from the inlet throughout the length of the pipes, without heat exchangers or intermediate working fluids.

The steam produced is then fed directly to the field's existing steam distribution network, where the steam is continuously injected deep into the oil reservoir. Sheltering the mirrors from the wind allows them to achieve higher temperature rates and prevents dust from building up as a result from exposure to humidity. GlassPoint Solar, the company that created the Enclosed Trough design, states its technology can produce heat for EOR for about $5 per million British thermal units in sunny regions, compared to between $10 and $12 for other conventional solar thermal technologies.

Power Tower Designs

Ivanpah Solar Electric Generating System with all three towers under load, Feb., 2014. Taken from I-15 in San Bernardino County, California. The Clark Mountain Range can be seen in the distance.

Power towers (also known as 'central tower' power plants or 'heliostat' power plants) capture and focus the sun's thermal energy with thousands of tracking mirrors (called heliostats) in roughly a two square mile field. A tower resides in the center of the heliostat field. The heliostats focus concentrated sunlight on a receiver which sits on top of the tower. Within the receiver the concentrated sunlight heats molten salt to over 1,000 °F (538 °C). The heated molten salt then flows into a thermal storage tank where it is stored, maintaining 98% thermal efficiency, and eventually pumped to a steam generator. The steam drives a standard turbine to generate electricity. This process, also known as the "Rankine cycle" is similar to a standard coal-fired power plant, except it is fueled by clean and free solar energy.

The advantage of this design above the parabolic trough design is the higher temperature. Thermal energy at higher temperatures can be converted to electricity more efficiently and can be more cheaply stored for later use. Furthermore, there is less need to flatten the ground area. In principle a power tower can be built on the side of a hill. Mirrors can be flat and plumbing is concentrated in the tower. The disadvantage is that each mirror must have its own dual-axis control, while in the parabolic trough design single axis tracking can be shared for a large array of mirrors.

A cost/performance comparison between power tower and parabolic trough concentrators was made by the NREL which estimated that by 2020 electricity could be produced from power towers for 5.47 ¢/kWh and for 6.21 ¢/kWh from parabolic troughs. The capacity factor for power towers was estimated to be 72.9% and 56.2% for parabolic troughs. There is some hope that the development of cheap, durable, mass producible heliostat power plant components could bring this cost down.

The first commercial tower power plant was PS10 in Spain with a capacity of 11 MW, completed in 2007. Since then a number of plants have been proposed, several have been built on a number of countries (Spain, Germany, U.S., Turkey, China, India) but several proposed plants were cancelled as photovoltaic solar prices plummeted. A solar power tower is expected to come online in South Africa in 2014. Ivanpah Solar Power Facility in California generates 392 MW of electricity from three towers, making it the largest solar power tower plant when it came online in late 2013.

Dish Designs

A parabolic solar dish concentrating the sun's rays on the heating element of a Stirling engine. The entire unit acts as a solar tracker.

CSP-Stirling is known to have the highest efficiency of all solar technologies (around 30%, compared to solar photovoltaic's approximately 15%), and is predicted to be able to produce the cheapest energy among all renewable energy sources in high-scale production and hot areas, semi-deserts, etc. A dish Stirling system uses a large, reflective, parabolic dish (similar in shape to a satellite television dish). It focuses all the sunlight that strikes the dish up onto a single point above the dish, where a receiver captures the heat and transforms it into a useful form. Typically the dish is coupled with a Stirling engine in a Dish-Stirling System, but also sometimes a steam engine is used. These create rotational kinetic energy that can be converted to electricity using an electric generator.

In 2005 Southern California Edison announced an agreement to purchase solar powered Stirling engines from Stirling Energy Systems over a twenty-year period and in quantities (20,000 units) sufficient to generate 500 megawatts of electricity. In January 2010, Stirling Energy Systems and Tessera Solar commissioned the first demonstration 1.5-megawatt power plant ("Maricopa Solar") using Stirling technology in Peoria, Arizona. At the beginning of 2011 Stirling Energy's development arm, Tessera Solar, sold off its two large projects, the 709 MW Imperial project and the 850 MW Calico project to AES Solar and K.Road, respectively. In 2012 the Maricopa plant was bought and dismantled by United Sun Systems. United Sun Systems released a new generation system, based on a V-shaped Stirling engine and a peak production of 33 kW. The new CSP-Stirling technology brings down LCOE to USD 0.02 in utility scale.

According to its developer, Rispasso Energy, a Swedish firm, in 2015 its Dish Sterling system being tested in the Kalahari Desert in South Africa showed 34% efficiency.

Fresnel Technologies

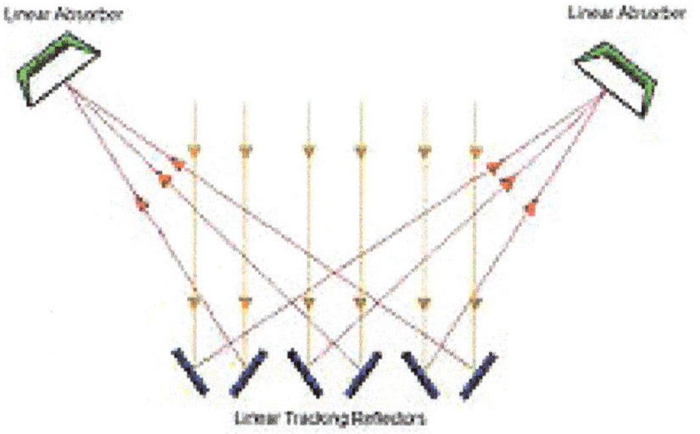

Fresnel reflector

A linear Fresnel reflector power plant uses a series of long, narrow, shallow-curvature (or even flat) mirrors to focus light onto one or more linear receivers positioned above the mirrors. On top of the receiver a small parabolic mirror can be attached for further focusing the light. These systems aim to offer lower overall costs by sharing a receiver between several mirrors (as compared with trough and dish concepts), while still using the simple line-focus geometry with one axis for tracking. This is similar to the trough design (and different from central towers and dishes with dual-axis). The receiver is stationary and so fluid couplings are not required (as in troughs and dishes). The mirrors also do not need to support the receiver, so they are structurally simpler. When suitable aiming strategies are used (mirrors aimed at different receivers at different times of day), this can allow a denser packing of mirrors on available land area.

Rival single axis tracking technologies include the relatively new linear Fresnel reflector (LFR) and compact-LFR (CLFR) technologies. The LFR differs from that of the parabolic trough in that the absorber is fixed in space above the mirror field. Also, the reflector is composed of many low row segments, which focus collectively on an elevated long tower receiver running parallel to the reflector rotational axis.

Prototypes of Fresnel lens concentrators have been produced for the collection of thermal energy by International Automated Systems. No full-scale thermal systems using Fresnel lenses are known to be in operation, although products incorporating Fresnel lenses in conjunction with photovoltaic cells are already available.

MicroCSP

MicroCSP is used for community-sized power plants (1 MW to 50 MW), for industrial, agricultural and manufacturing 'process heat' applications, and when large amounts of hot water are needed, such as resort swimming pools, water parks, large laundry facilities, sterilization, distillation and other such uses.

Enclosed Parabolic Trough

The enclosed parabolic trough solar thermal system encapsulates the components within an off-the-shelf greenhouse type of glasshouse. The glasshouse protects the components from the elements that can negatively impact system reliability and efficiency. This protection importantly includes nightly glass-roof washing with optimized water-efficient off-the-shelf automated washing systems. Lightweight curved solar-reflecting mirrors are suspended from the ceiling of the glasshouse by wires. A single-axis tracking system positions the mirrors to retrieve the optimal amount of sunlight. The mirrors concentrate the sunlight and focus it on a network of stationary steel pipes, also suspended from the glasshouse structure. Water is pumped through the pipes and boiled to generate steam when intense sun radiation is applied. The steam is available for process heat. Sheltering the mirrors from the wind allows them to achieve higher temperature rates and prevents dust from building up on the mirrors as a result from exposure to humidity.

Heat Collection and Exchange

More energy is contained in higher frequency light based upon the formula of $E = h v$, where h is the Planck constant and v is frequency. Metal collectors down convert higher frequency light by producing a series of Compton shifts into an abundance of lower frequency light. Glass or ceramic coatings with high transmission in the visible and UV and effective absorption in the IR (heat blocking) trap metal absorbed low frequency light from radiation loss. Convection insulation prevents mechanical losses transferred through gas. Once collected as heat, thermos containment efficiency improves significantly with increased size. Unlike Photovoltaic technologies that often degrade under concentrated light, Solar Thermal depends upon light concentration that requires a clear sky to reach suitable temperatures.

Heat in a solar thermal system is guided by five basic principles: heat gain; heat transfer; heat storage; heat transport; and heat insulation. Here, heat is the measure of the amount of thermal energy an object contains and is determined by the temperature, mass and specific heat of the object. Solar thermal power plants use heat exchangers that are designed for constant working conditions, to provide heat exchange. Copper heat exchangers are important in solar thermal heating and cooling systems because of copper's high thermal conductivity, resistance to atmospheric and water corrosion, sealing and joining by soldering, and mechanical strength. Copper is used both in receivers and in primary circuits (pipes and heat exchangers for water tanks) of solar thermal water systems.

Heat gain is the heat accumulated from the sun in the system. Solar thermal heat is trapped using the greenhouse effect; the greenhouse effect in this case is the ability of a reflective surface to transmit short wave radiation and reflect long wave radiation. Heat and infrared radiation (IR) are produced when short wave radiation light hits the absorber plate, which is then trapped inside the collector. Fluid, usually water, in the absorber tubes collect the trapped heat and transfer it to a heat storage vault.

Heat is transferred either by conduction or convection. When water is heated, kinetic energy is transferred by conduction to water molecules throughout the medium. These molecules spread their thermal energy by conduction and occupy more space than the cold slow moving molecules above them. The distribution of energy from the rising hot water to the sinking cold water contributes to the convection process. Heat is transferred from the absorber plates of the collector in the fluid by conduction. The collector fluid is circulated through the carrier pipes to the heat transfer vault. Inside the vault, heat is transferred throughout the medium through convection.

Heat storage enables solar thermal plants to produce electricity during hours without sunlight. Heat is transferred to a thermal storage medium in an insulated reservoir during hours with sunlight, and is withdrawn for power generation during hours lacking sunlight. Thermal storage mediums will be discussed in a heat storage section. Rate of heat transfer is related to the conductive and convection medium as well as the temperature differences. Bodies with large temperature differences transfer heat faster than bodies with lower temperature differences.

Heat transport refers to the activity in which heat from a solar collector is transported to the heat storage vault. Heat insulation is vital in both heat transport tubing as well as the storage vault. It prevents heat loss, which in turn relates to energy loss, or decrease in the efficiency of the system.

Heat Storage for Space Heating

A collection of mature technologies called seasonal thermal energy storage (STES) is capable of storing heat for months at a time, so solar heat collected primarily in Summer can be used for all-year heating. Solar-supplied STES technology has been advanced primarily in Denmark, Germany, and Canada, and applications include individual buildings and district heating networks. Drake Landing Solar Community in Alberta, Canada has a small district system and in 2012 achieved a world record of providing 97% of the community's all-year space heating needs from the sun. STES thermal storage mediums include deep aquifers; native rock surrounding clusters of small-diameter, heat exchanger equipped boreholes; large, shallow, lined pits that are filled with gravel and top-insulated; and large, insulated and buried surface water tanks.

Heat Storage to Stabilize Solar-electric Power Generation

Heat storage allows a solar thermal plant to produce electricity at night and on overcast days. This allows the use of solar power for baseload generation as well as peak power generation, with the potential of displacing both coal- and natural gas-fired power plants. Additionally, the utilization of the generator is higher which reduces cost. Even short term storage can help by smoothing out the "duck curve" of rapid change in generation requirements at sunset when a grid includes large amounts of solar capacity.

Heat is transferred to a thermal storage medium in an insulated reservoir during the day, and withdrawn for power generation at night. Thermal storage media include pressurized steam, concrete, a variety of phase change materials, and molten salts such as calcium, sodium and potassium nitrate.

Steam Accumulator

The PS10 solar power tower stores heat in tanks as pressurized steam at 50 bar and 285 °C. The steam condenses and flashes back to steam, when pressure is lowered. Storage is for one hour. It is suggested that longer storage is possible, but that has not been proven yet in an existing power plant.

Molten Salt Storage

The 150 MW Andasol solar power station is a commercial parabolic trough solar thermal power plant, located in Spain. The Andasol plant uses tanks of molten salt to store solar energy so that it can continue generating electricity even when the sun isn't shining.

A variety of fluids have been tested to transport the sun's heat, including water, air, oil, and sodium, but Rockwell International selected molten salt as best. Molten salt is used in solar power tower systems because it is liquid at atmospheric pressure, provides a low-cost medium to store thermal energy, its operating temperatures are compatible with today's steam turbines, and it is non-flammable and nontoxic. Molten salt is used in the chemical and metals industries to transport heat, so industry has experience with it.

The first commercial molten salt mixture was a common form of saltpeter, 60% sodium nitrate and 40% potassium nitrate. Saltpeter melts at 220 °C (430 °F) and is kept liquid at 290 °C (550 °F) in an insulated storage tank. Calcium nitrate can reduce the melting point to 131 °C, permitting more energy to be extracted before the salt freezes. There are now several technical calcium nitrate grades stable at more than 500 °C.

This solar power system can generate power in cloudy weather or at night using the heat in the tank of hot salt. The tanks are insulated, able to store heat for a week. Tanks that power a 100-megawatt turbine for four hours would be about 9 m (30 ft) tall and 24 m (80 ft) in diameter.

The Andasol power plant in Spain is the first commercial solar thermal power plant using molten salt for heat storage and nighttime generation. It came on line March 2009. On July 4, 2011, a company in Spain celebrated an historic moment for the solar industry: Torresol's 19.9 MW concentrating solar power plant became the first ever to generate uninterrupted electricity for 24 hours straight, using a molten salt heat storage.

In 2016 SolarReserve proposed a 2 GW, $5 billion concentrated solar plant with storage in Nevada.

Phase-change Materials for Storage

Phase Change Material (PCMs) offer an alternative solution in energy storage. Using a similar heat transfer infrastructure, PCMs have the potential of providing a more efficient means of storage. PCMs can be either organic or inorganic materials. Advantages of organic PCMs include no corrosives, low or no undercooling, and chemical and thermal stability. Disadvantages include low phase-change enthalpy, low thermal conductivity, and flammability. Inorganics are advantageous with greater phase-change enthalpy, but exhibit disadvantages with undercooling, corrosion, phase separation, and lack of thermal stability. The greater phase-change enthalpy in inorganic PCMs make hydrate salts a strong candidate in the solar energy storage field.

Use of Water

A design which requires water for condensation or cooling may conflict with location of solar thermal plants in desert areas with good solar radiation but limited water resources. The conflict is illustrated by plans of Solar Millennium, a German company, to build a plant in the Amargosa Valley of Nevada which would require 20% of the water available in the area. Some other projected plants by the same and other companies in the Mojave Desert of California may also be affected by difficulty in obtaining adequate and appropriate water rights. California water law currently prohibits use of potable water for cooling.

Other designs require less water. The Ivanpah Solar Power Facility in south-eastern California conserves scarce desert water by using air-cooling to convert the steam back into water. Compared to conventional wet-cooling, this results in a 90% reduction in water usage at the cost of some loss of efficiency. The water is then returned to the boiler in a closed process which is environmentally friendly.

Conversion Rates from Solar Energy to Electrical Energy

Of all of these technologies the solar dish/Stirling engine has the highest energy efficiency. A single solar dish-Stirling engine installed at Sandia National Laboratories National Solar Thermal Test Facility (NSTTF) produces as much as 25 kW of electricity, with a conversion efficiency of 31.25%.

Solar parabolic trough plants have been built with efficiencies of about 20%. Fresnel reflectors have an efficiency that is slightly lower (but this is compensated by the denser packing).

The gross conversion efficiencies (taking into account that the solar dishes or troughs occupy only a fraction of the total area of the power plant) are determined by net generating capacity over the solar energy that falls on the total area of the solar plant. The 500-megawatt (MW) SCE/SES plant would extract about 2.75% of the radiation that falls on its 4,500 acres (18.2 km^2). For the 50 MW AndaSol Power Plant that is being built in Spain (total area of $1{,}300 \times 1{,}500$ m = 1.95 km^2) gross conversion efficiency comes out at 2.6%.

Furthermore, efficiency does not directly relate to cost: on calculating total cost, both efficiency and the cost of construction and maintenance should be taken into account.

Solar Thermal Power Plants

The two main types of solar thermal power plants are

1. Concentrating Solar Power (CSP) plants.
2. Solar Chimneys

Concentrating Solar Power (CSP) plants

Solar thermal power plants generally use reflectors to concentrate sunlight into a heat absorber. Such power plants are known as Concentrating Solar Power (CSP) plants.

Concentrating solar power plants produce electric power by converting the sun's energy into high-temperature heat using various mirror configurations. The heat is then channeled through a conventional generator. The plants consist of two parts, one that collects solar energy and converts it to heat, and another that converts heat energy to electricity.

Concentrating solar power systems can be sized for village power (10 kilowatts) or grid- connected applications (up to 100 megawatts). Some systems use thermal storage during cloudy periods or at night. There are four CSP technologies being promoted internationally. For each of these, there exists various design variations or different configurations.

The amount of power generated by a concentrating solar power plant depends on the amount of direct sunlight. Like concentrating photovoltaic concentrators, these technologies use only direct-beam sunlight, rather than diffuse solar radiation.

Types of CSP plants

A. Parabolic Trough Systems:

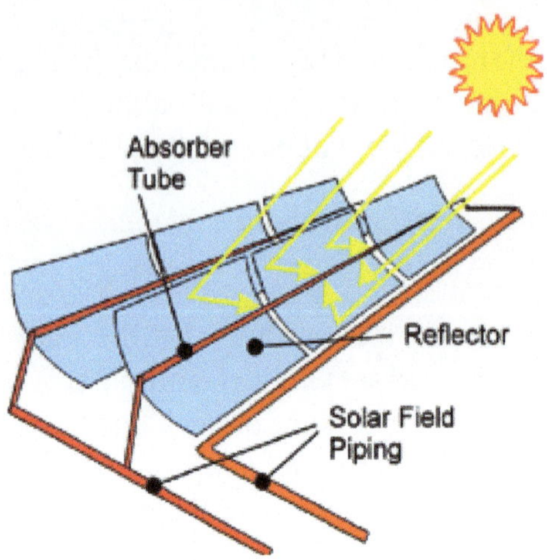

The sun's energy is concentrated by parabolically curved, trough-shaped reflectors onto a receiver pipe running along the inside of the curved surface. This energy heats oil flowing through the pipe and the heat energy is then used to generate electricity in a conventional steam generator.

A collector field comprises many troughs in parallel rows aligned on a north-south axis. This configuration enables the single-axis troughs to track the sun from east to west during the day to ensure that the sun is continuously focused on the receiver pipes. Individual trough systems currently can generate about 80 megawatts of electricity.

Trough designs can incorporate thermal storage—setting aside the heat transfer fluid in its hot phase—allowing for electricity generation several hours into the evening.

Currently, all parabolic trough plants are "hybrids," meaning they use fossil fuel to supplement the solar output during periods of low solar radiation.

Another option under investigation is the approximation of the parabolic troughs by segmented mirrors according to the principle of Fresnel.

B. Power Tower Systems:

A power tower converts sunshine into clean electricity for the electricity grids. The technology utilizes many large, sun-tracking mirrors (heliostats) to focus sunlight on a receiver at the top of a tower. A heat transfer fluid heated in the receiver is used to generate steam, which, in turn, is used in a conventional turbine-generator to produce electricity.

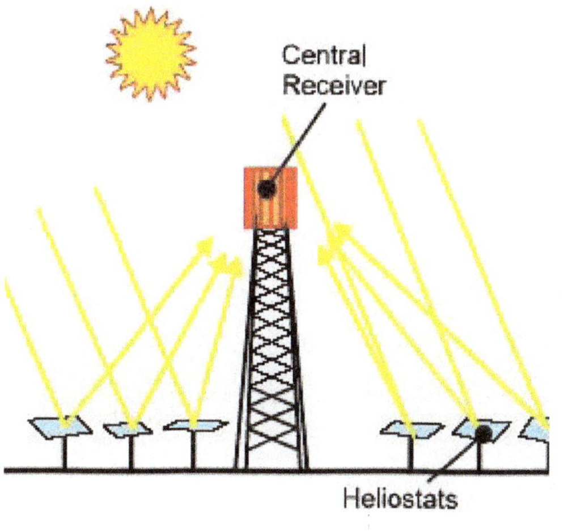

Early power towers (such as the Solar One plant) utilized steam as the heat transfer fluid; current designs (including Solar Two, shown in fig) utilize molten nitrate salt because of its superior heat transfer and energy storage capabilities. Current European designs use air as heat transfer medium because of its high temperature and its good hand ability.

Individual commercial plants will be sized to produce anywhere from 50 to 200 MW of electricity.

C. Parabolic Dish Systems:

Parabolic dish systems consist of a parabolic-shaped point focus concentrator in the form of a dish that reflects solar radiation onto a receiver mounted at the focal point. These concentrators are mounted on a structure with a two-axis tracking system to follow the sun.

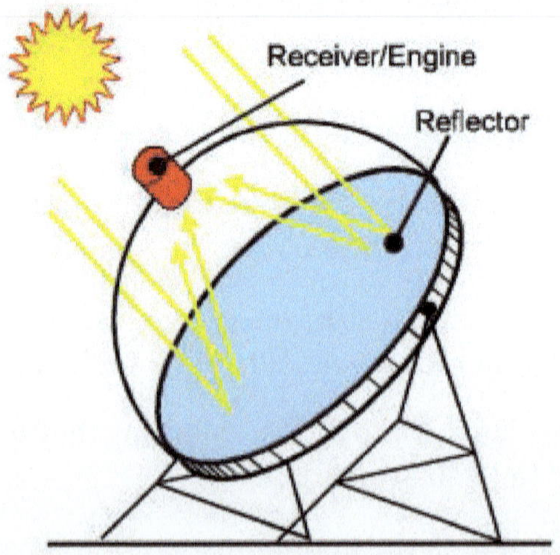

The collected heat is typically utilized directly by a heat engine mounted on the receiver moving with the dish structure. Stirling and Brayton cycle engines are currently favored for power conversion.

Projects of modular systems have been realized with total capacities up to 5 MW. The modules have max sizes of 50 kW and have achieved peak efficiencies up to 30% net.

Solar Chimney

A solar chimney is a solar thermal power plant where air passes under a very large agricultural glass house (between 2 and 30 kilometers in diameter); the air is heated by the sun and channeled upwards towards a convection tower. It then rises naturally and is used to drive turbines, which generate electricity.

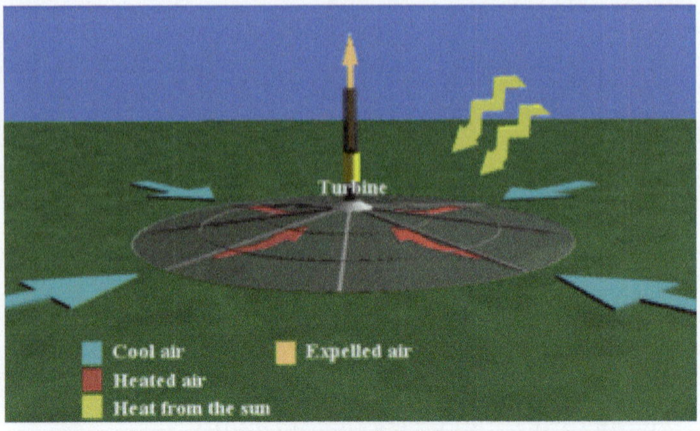

General concept of proposed solar chimney power station

A solar chimney is an days such a chimney can provide ventilation where there apparatus for harnessing solar energy by convection of heated air. In its simplest form, it simply consists of a black-painted chimney. During the daytime, solar energy heats the chimney and thereby heats the air within it, resulting in an Updraft of air within the chimney. The suction this creates at the chimney base can also be used to ventilate, and thereby cool, the building below. In most parts of the world, it is easier to harness wind power for such ventilation, but on hot windless would otherwise be none. This principle has been proposed for electric power generation, using a large greenhouse at the base rather than relying on heating of the chimney itself. The main problem with this approach is the relatively small difference in temperature between the highest and lowest temperatures in the system. Carnot's theorem greatly restricts the efficiency of conversion in these circumstances.

Water Heating

Water heating is required in most countries of the world for both domestic and commercial use. There are a wide variety of solar water heaters available. The simplest solar water heater is a piece of black plastic pipe, filled with water, and laid in the sun for the water to heat up. Simple solar water heaters usually comprise a series of pipes, which are painted black, sitting inside an insulated box fronted with a glass panel. This is known as a solar collector. The fluid to be heated passes through the collector and into a tank for storage. The fluid can be cycled through the tank several times to raise the heat of the fluid to the required temperature. There are two common simple configurations for such a system and they are outlined below.

a. The thermosyphon system makes use of the natural tendency of hot water to rise above cold water. The tank in such a system is always placed above the top of the collector and as water is heated in the collector it rises and is replaced by cold water from the bottom of the tank. This cycle will continue until the temperature of the water in the tank is equal to that of the panel. A one-way valve is usually fitted in the system to prevent the reverse occurring at night when the temperature drops. As hot water is drawn off for use, fresh cold water is fed into the system from the mains. As most solar collectors are fitted on the roofs of houses, this system is not always convenient, as it is difficult to site the tank above the collector, in which case the system will need A pump to circulate the water.

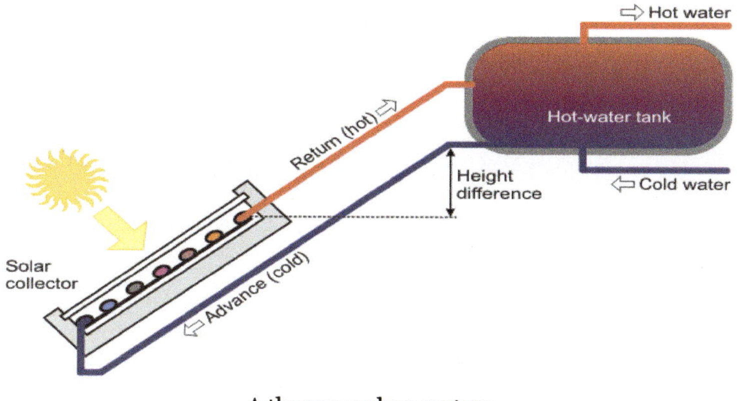

A thermosyphon system

b. Pumped solar water heaters use a pumping device to drive the water through the collector. The advantage of this system is that the storage tank can be sited below the collector. The disadvantage of course is that electricity is required to drive the pump. Often the fluid circulating in the collector will be treated with an anti-corrosive and /or anti-freeze chemical. In this case a heat exchanger is required to transfer the heat to the consumers hot water supply.

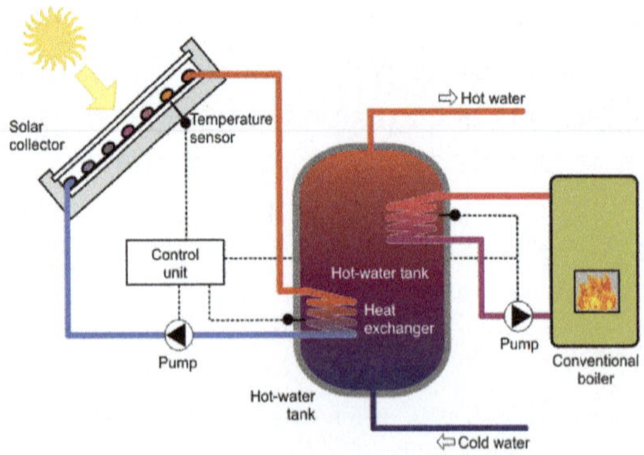

A double-cycle system with forced circulation with a conventional boiler for back-up heating

Solar District Heating

If an entire housing estate should be fitted with solar systems, one solution is a solar district heating system. The collectors are either distributed on the houses, or replaced by a large, central solar collector. The collectors then heat up a big central storage tank, from which much of the heat is distributed back to the houses. The surface-to-volume ratio of a central storage tank is much better than that for distributed storage systems, so the storage losses are much lower, and even permit seasonal heat storage. Solar district heating is also an option if room heating is to be covered by solar energy. There are higher piping losses with a central tank, but some solar district heating demonstration systems have already been successfully tested.

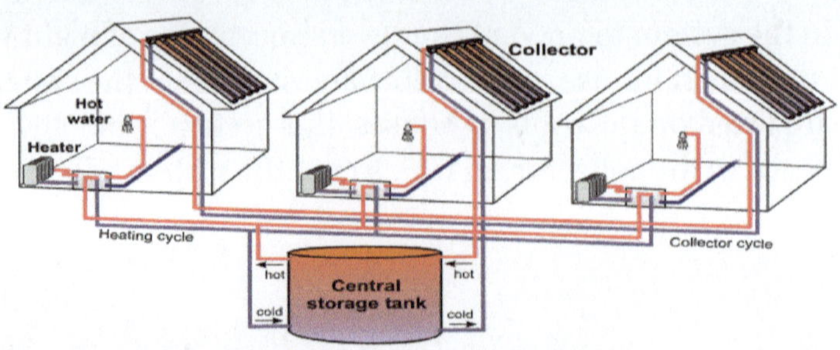

A solar district heating system

Cost Benefits of Solar Water Heating System

The most cost- effective way to install a solar geyser is to integrate the collector assembly, cold-water supply and piping with the design of a new house under construction. Solar geysers can easily be installed in group houses and apartments, especially during construction, if adequate provisions are made for piping, collector assembly and cold-water supply. Proper load matching is required to ensure that the capacity of the system installed is optimized to meet The daily hot water needs of the end-user.

Current prices of domestic SWHs are around Rest. 20,000 for a 100 litres per day system. The life-cycle cost of energy of saved electrical energy is Rest. 5600 per year at the point of end- user. If the SWH is purchased as an add-on to an existing geyser, the simple payback period for the consumer would be 5.45 years. If the SWH is purchased instead of a conventional geyser, the payback period for the consumer would reduce to 3.58 years.

Solar Dryer

Controlled drying is required for various crops and products, such as grain, coffee, tobacco, fruits vegetables and fish. Their quality can be enhanced if the drying is properly carried out. Solar thermal technology can be used to assist with the drying of such products. Solar drying is in practice since the time imp-memorable for preservation of food and agriculture crops. This was done particularly by open sun drying under open the sky.

In open air Solar drying the heat is supplied by direct absorption of solar radiation by material being dried. The vapor produced is carried away by air moving past the material, the air motion being due either to natural convection resulting from contact with the heated material or to winds.

This process has several disadvantages.

Disadvantages of Mechanical and Artificial Drying:

1. Spoilage of product due to adverse climatic condition like rain, wind etc
2. Loss of material due to birds and animals
3. Deterioration of the material by decomposition, insects and fungus growth
4. Highly energy intensive and expensive

Solar dryer make use of solar radiation, ambient temperature, relative humidity. Heated air is passed naturally or mechanically circulated to remove moisture from material placed in side the enclosure.

Solar dryer is a very useful device for:

1. Agriculture crop drying
2. Food processing industries for dehydration of fruits, potatoes, onions and other vegetables,
3. Dairy industries for production of milk powder, casein etc.
4. Seasoning of wood and timber.
5. Textile industries for drying of textile materials.

Working of Solar Dryer

The main principle of operation is to raise the heat of the product, which is usually held within a compartment or box, while at the same time passing air through the compartment to remove moisture. The flow of air is often promoted using the 'stack' effect which takes advantage of the fact that hot air rises and can therefore be drawn upwards through a chimney, while drawing in Cooler air from below. Alternatively a fan can be used. The size and shape of the compartment Varies depending on the product and the scale of the drying system. Large systems can use Large barns while smaller systems may have a few trays in a small wooden housing.

Solar crop drying technologies can help reduce environmental degradation caused by the use Of fuel wood or fossil fuels for crop drying and can also help to reduce the costs associated.

With these fuels and hence the cost of the product.

The principal types of solar dryers are enumerated below.

1. Solar cabinet dryer
2. Solar green house dryers
3. Indirects sonars drayer

Solar Stills

Solar still is a device to desalinate impure water like brackish or saline water. It a simple device to get potable/fresh distilled water from impure water, using solar energy as fuel, for its various applications in domestic, industrial and academic sectors. A solar still consist of shallow triangular basin made up of Fiber Reinforced Plastic (FRP). Bottom of the basin is painted black so as to absorb solar heat effectively. Top of the basin is covered with transparent glass tilt fitted so that maximum solar radiation can be transmitted in to the still. Edges of the glass are sealed with the basin using tar tape so that the entire basin becomes airtight. Entire assembly is placed on a structure made of MS angle. Out let is connected with a storage container. Provision has been made to fill water in the still basin. A window is provided in the basin to clean the basin from inside. Water is charged in to the basin in a thin layer.

Solar Stills have got major advantages over other conventional Distillation / water purification / de- mineralization systems as follows:

1. Produces pure water
2. No prime movers required
3. No conventional energy required
4. No skilled operator required
5. Local manufacturing/repairing
6. Low investment
7. Can purify highly saline water (even sea water)

Working of Solar Still

Working of solar still is based on simple scientific principle of Evaporation and condensation. Brackish or saline water is filled in still basin, which is painted black at the bottom. Solar radiation received at the surface is absorbed effectively by the blacken surface and heat is transferred to the water in the basin. Temperature of the water increases that increases rate of evaporation. Water vapour formed by evaporation rises upward and condenses on the inner surface of the cover glass, which is relatively cold.

Condensed water vapor trickles down in to troughs from there it is collected in to the storage container.

The operation of Solar Still is very simple and no special skill is required for its operation and maintenance. Solar stills is an useful devise to get fresh/ distilled water which is required in Industries for industrial processes Hospitals and Dispensaries for sterilization Garages and Automobile Workshop for radiator and battery maintenance Telephone Exchange for battery maintenance Laboratory Use for analytic work Marshy and costal area To get fresh potable water.

Solar Box Cooker

A solar box cooker is an insulated transparent-topped box with a reflective lid. It is designed to capture solar power and keep its interior warm. The major parts of a solar cooker are enumerated below.

Important Parts of Solar Cooker

1. Outer Box: The outer box of a solar cooker is generally made of G.I. or aluminum sheet or fiber reinforced plastic.

2. Inner Cooking Box (Tray): This is made from aluminum sheet. The inner cooking box is slightly smaller than the outer box. It is coated with black paint so as to easily absorb solar radiation and transfer the heat to the cooking pots.

3. Double Glass Lid: A double glass lid covers the inner box or tray. This cover is slightly larger than the inner box. The two glass sheets are fixed in an aluminum frame with a spacing of 2 centimeters between the two glasses. This space contains air which insulates and prevents heat escaping from inside. A rubber strip is affixed on the edges of the frame to prevent any heat leakage.

4. Thermal Insulator: The space between the outer box and inner tray including bottom of the tray is packed with insulating material such as glass wool pads to reduce heat losses from the cooker. This insulating material should be free from volatile materials.

5. Mirror: Mirror is used in a solar cooker to increase the radiation input on the absorbing space and is fixed on the inner side of the main cover of the box. Sunlight falling on the mirror gets reflected from it and enters into the tray through the double glass lid. This radiation is in addition to the radiation entering the box directly and helps to quicken the cooking process by raising the inside temperature of the cooker.

6. Containers: The cooking containers (with cover) are generally made of aluminum or stainless steel. These pots are also painted black on the outer surface so that they also absorb solar radiation directly.

The solar box cooker typically reaches a temperature of 90 °C; not as hot as a standard oven, but still enough to warm food over an hour. Because it doesn't reach too high a temperature, food can be safely left in it all day without burning. The cooker is often used to make a large pot of food in the morning, and then people eat servings or snack from it all day. The cooker is usually used to warm food and drinks but can also be used to pasteurize milk and sanitize water.

Horace de Assure, a Swiss naturalist, invented solar cookers as early as 1767

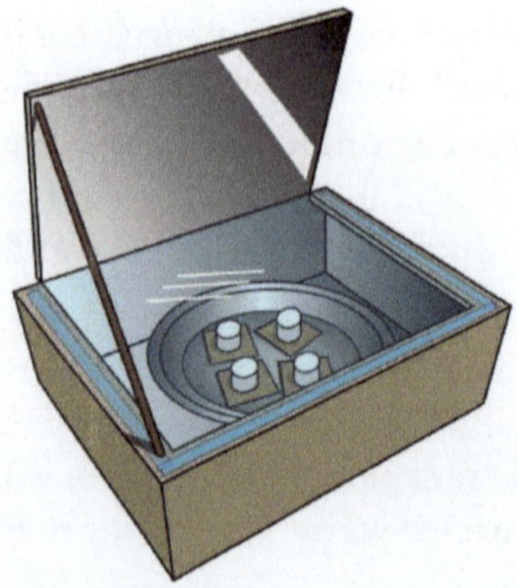

We can cook a large number of items, like pulses, rice, cheer etc. The time taken to cook will depend upon the type of the food, time of the day and solar intensity. However, the time taken to cook some of the dishes in a solar cooker is as follows:

1. Rice (45 minutes to one hour),
2. Vegetables (about one to two hours),
3. Black gram and Raja (about two hours),
4. Cake (one hour).

Community Solar Cooker

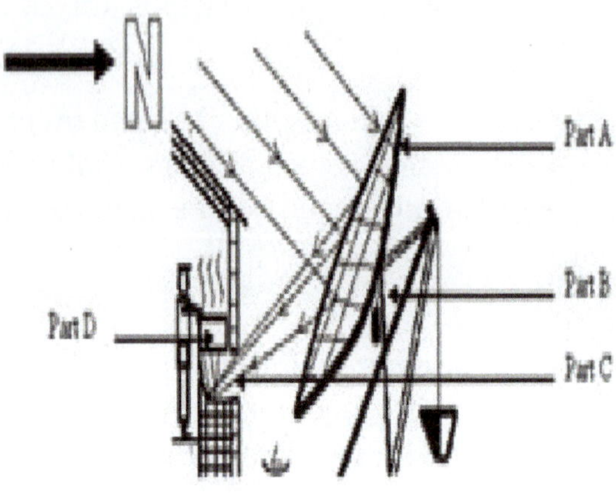

Numerous households all over the country are using 'Surry' Cooker to cook their meals. This family-sized cooker can cook meals for 4-5 persons. A larger version of the family size box-type cooker was also developed and used for canteen application. The canteen size solar cookers are just larger in size and can cook for 10-15 persons. For meeting still larger demands a large Community Solar Cooker has been developed with Solar Concentrator technology. It is an ideal cooking device for hostels, guesthouses etc.

Firewood is the most commonly used cooking fuel in community kitchens and traditional wood-stove - a "Chula" are the most commonly used cooking device. These chelas have an efficiency of 5-10 % only. The implications of utilizing such inefficient devices are higher fuel consumption & environment pollution. As a result more trees are felled each year for meeting the ever-increasing energy demands. This leads to rapid deforestation and environment degradation. In addition, the smoke from chelas causes serious health hazard to kitchen workers. The Community Solar Cooker, in contrast, poses no such health hazards as it utilizes 'the freely available eco-friendly solar energy from nature.

This Community Solar Cooker employs a parabolic reflecting concentrator that can cook large quantities of food at much faster rate. The best part is that the cooking can take place within an Enclosed kitchen. It can replace LPG, kerosene and firewood which are either cumbersome to use, very expensive or which are in short supply. This Cooker is technologically far superior to the box-type community cooker, having overcome a number of shortcomings of the box type cooker.

This cooker is capable of achieving higher temperature unto $250°C$ as against $100-125°C$ in box type cooker. This helps cooking much faster. The conventional cooking arrangement within the kitchen does not require to be changed and the cooking can be done inside the kitchen. Additionally roasting & frying can be done with this cooker, which is not possible in the old box type solar cooker.

The cost of Cooker, inclusive of all attachments and installation charges is about Rest. 55,000.

Solar Air Conditioning

The basic principle behind solar thermal driven cooling is the thermo-chemical process of Sorption: a liquid or gaseous substance is either attached to a solid, porous material called Adsorption or is taken in by a liquid or solid material called Absorption.

The sorbet, silica gel, a substance with a large inner surface area is provided with heat From a solar heater and is dehumidified. After this "drying", the process can be repeated in the opposite direction. When providing water vapor or steam, it is stored in the porous storage medium (adsorption) and simultaneously heat is released.

Processes are differentiated between closed refrigerant circulation systems, for producing Cold water, and open systems according to the way in which the process is carried out. That Is, whether or not the refrigerant comes into contact with the atmosphere. The latter is used For dehumidification and evaporative cooling.

Basic structure of a solar air conditioning system:

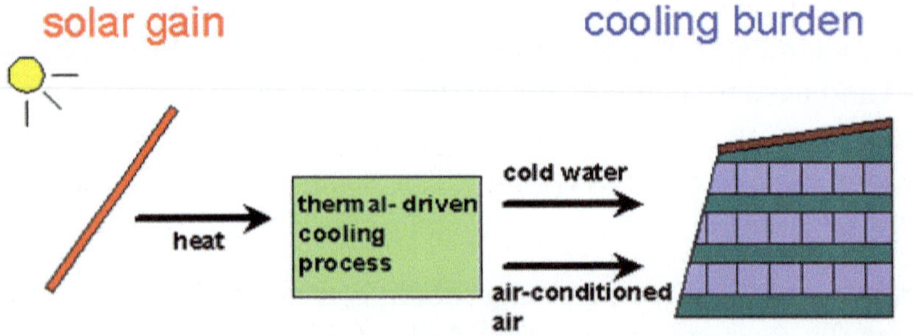

Proposed Application

A solar power satellite, or SPS, is a satellite built in high Earth orbit that uses microwave power Transmission to beam solar power to a very large antenna on Earth where it can be used in place of conventional power sources. The advantage to placing the solar collectors in space is the unobstructed view of the Sun, unaffected by the day/night cycle, weather, or seasons.

However, the costs of construction are very high, so it is unlikely the SPS will be able to compete with conventional sources unless there is a big reduction in the costs associated with launching massive satellites into space, unless a space-based manufacturing industry develops and they can be built in orbit.

Another problem is the dangers involved in transmitting such high power radiations through the atmosphere.

Solar Power

Solar power is the conversion of energy from sunlight into electricity, either directly using photovoltaics (PV), or indirectly using concentrated solar power. Concentrated solar power systems use lenses or mirrors and tracking systems to focus a large area of sunlight into a small beam. Photovoltaic cells convert light into an electric current using the photovoltaic effect.

A solar photovoltaic system array on a rooftop in Hong Kong

Photovoltaics were initially solely used as a source of electricity for small and medium-sized applications, from the calculator powered by a single solar cell to remote homes powered by an off-grid rooftop PV system. Commercial concentrated solar power plants were first developed in the 1980s. The 392 MW Ivanpah installation is the largest concentrating solar power plant in the world, located in the Mojave Desert of California.

The first three concentrated solar power (CSP) units of Spain's Solnova Solar Power Station in the foreground, with the PS10 and PS20 solar power towers in the background

As the cost of solar electricity has fallen, the number of grid-connected solar PV systems has grown into the millions and utility-scale solar power stations with hundreds of megawatts are being built. Solar PV is rapidly becoming an inexpensive, low-carbon technology to harness renewable energy from the Sun. The current largest photovoltaic power station in the world is the 850 MW Longyangxia Dam Solar Park, in Qinghai, China.

The International Energy Agency projected in 2014 that under its "high renewables" scenario, by 2050, solar photovoltaics and concentrated solar power would contribute about 16 and 11 percent, respectively, of the worldwide electricity consumption, and solar would be the world's largest source of electricity. Most solar installations would be in China and India. Currently, as of 2016, solar power provides just 1% of total worldwide electricity production but growing 33% per annum.

Mainstream Technologies

Many industrialized nations have installed significant solar power capacity into their grids to supplement or provide an alternative to conventional energy sources while an increasing number of less developed nations have turned to solar to reduce dependence on expensive imported fuels. Long distance transmission allows remote renewable energy resources to displace fossil fuel consumption. Solar power plants use one of two technologies:

- Photovoltaic (PV) systems use solar panels, either on rooftops or in ground-mounted solar farms, converting sunlight directly into electric power.
- Concentrated solar power (CSP, also known as "concentrated solar thermal") plants use solar thermal energy to make steam, that is thereafter converted into electricity by a turbine.

Photovoltaics

A solar cell, or photovoltaic cell (PV), is a device that converts light into electric current using the photovoltaic effect. The first solar cell was constructed by Charles Fritts in the 1880s. The German industrialist Ernst Werner von Siemens was among those who recognized the importance of this discovery. In 1931, the German engineer Bruno Lange developed a photo cell using silver selenide in place of copper oxide, although the prototype selenium cells converted less than 1% of incident light into electricity. Following the work of Russell Ohl in the 1940s, researchers Gerald Pearson, Calvin Fuller and Daryl Chapin created the silicon solar cell in 1954. These early solar cells cost 286 USD/watt and reached efficiencies of 4.5–6%.

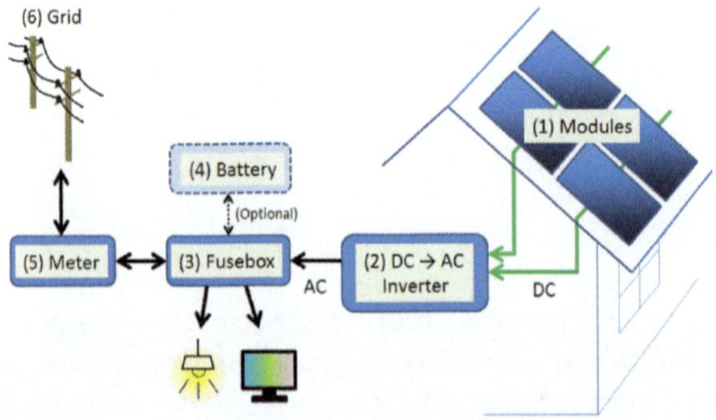

Schematics of a grid-connected residential PV power system

Conventional PV Systems

The array of a photovoltaic power system, or PV system, produces direct current (DC) power which fluctuates with the sunlight's intensity. For practical use this usually requires conversion to certain desired voltages or alternating current (AC), through the use of inverters. Multiple solar cells are connected inside modules. Modules are wired together to form arrays, then tied to an inverter, which produces power at the desired voltage, and for AC, the desired frequency/phase.

Many residential PV systems are connected to the grid wherever available, especially in developed countries with large markets. In these grid-connected PV systems, use of energy storage is optional. In certain applications such as satellites, lighthouses, or in developing countries, batteries or additional power generators are often added as back-ups. Such stand-alone power systems permit operations at night and at other times of limited sunlight.

Concentrated Solar Power

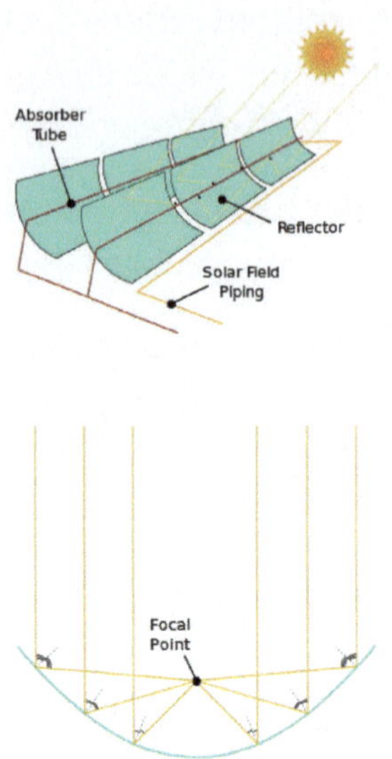

A parabolic collector concentrates sunlight onto a tube in its focal point.

Concentrated solar power (CSP), also called "concentrated solar thermal", uses lenses or mirrors and tracking systems to focus a large area of sunlight into a small beam. Contrary to photovoltaics – which converts light directly into electricity – CSP uses the heat of the sun's radiation to generate electricity from conventional steam-driven turbines.

A wide range of concentrating technologies exists: among the best known are the parabolic trough, the compact linear Fresnel reflector, the Stirling dish and the solar power tower. Various techniques are used to track the sun and focus light. In all of these systems a working fluid is heated by the concentrated sunlight, and is then used for power generation or energy storage. Thermal storage efficiently allows up to 24-hour electricity generation.

A *parabolic trough* consists of a linear parabolic reflector that concentrates light onto a receiver positioned along the reflector's focal line. The receiver is a tube positioned right above the middle of the parabolic mirror and is filled with a working fluid. The reflector is made to follow the sun during daylight hours by tracking along a single axis. Parabolic trough systems provide the best land-use factor of any solar technology. The SEGS plants in California and Acciona's Nevada Solar One near Boulder City, Nevada are representatives of this technology.

Compact Linear Fresnel Reflectors are CSP-plants which use many thin mirror strips instead of parabolic mirrors to concentrate sunlight onto two tubes with working fluid. This has the advantage that flat mirrors can be used which are much cheaper than parabolic mirrors, and that more reflectors can be placed in the same amount of space, allowing more of the available sunlight to be used. Concentrating linear fresnel reflectors can be used in either large or more compact plants.

The *Stirling solar dish* combines a parabolic concentrating dish with a Stirling engine which normally drives an electric generator. The advantages of Stirling solar over photovoltaic cells are higher efficiency of converting sunlight into electricity and longer lifetime. Parabolic dish systems give the highest efficiency among CSP technologies. The 50 kW Big Dish in Canberra, Australia is an example of this technology.

A *solar power tower* uses an array of tracking reflectors (heliostats) to concentrate light on a central receiver atop a tower. Power towers are more cost effective, offer higher efficiency and better energy storage capability among CSP technologies. The PS10 Solar Power Plant and PS20 solar power plant are examples of this technology.

Hybrid Systems

A hybrid system combines (C)PV and CSP with one another or with other forms of generation such as diesel, wind and biogas. The combined form of generation may enable the system to modulate power output as a function of demand or at least reduce the fluctuating nature of solar power and the consumption of non renewable fuel. Hybrid systems are most often found on islands.

CPV/CSP System

A novel solar CPV/CSP hybrid system has been proposed, combining concentrator photovoltaics with the non-PV technology of concentrated solar power, or also known as concentrated solar thermal.

ISCC System

The Hassi R'Mel power station in Algeria, is an example of combining CSP with a gas turbine, where a 25-megawatt CSP-parabolic trough array supplements a much larger 130 MW combined cycle gas turbine plant. Another example is the Yazd power station in Iran.

PVT System

Hybrid PV/T), also known as *photovoltaic thermal hybrid solar collectors* convert solar radiation into thermal and electrical energy. Such a system combines a solar (PV) module with a solar thermal collector in an complementary way.

CPVT System

A concentrated photovoltaic thermal hybrid (CPVT) system is similar to a PVT system. It uses concentrated photovoltaics (CPV) instead of conventional PV technology, and combines it with a solar thermal collector.

PV Diesel System

It combines a photovoltaic system with a diesel generator. Combinations with other renewables are possible and include wind turbines.

PV-thermoelectric System

Thermoelectric, or "thermovoltaic" devices convert a temperature difference between dissimilar materials into an electric current. Solar cells use only the high frequency part of the radiation, while the low frequency heat energy is wasted. Several patents about the use of thermoelectric devices in tandem with solar cells have been filed. The idea is to increase the efficiency of the combined solar/thermoelectric system to convert the solar radiation into useful electricity.

Development and Deployment

Electricity Generation from Solar		
Year	**Energy (TWh)**	**% of Total**
2004	2.6	0.01%
2005	3.7	0.02%
2006	5.0	0.03%
2007	6.8	0.03%
2008	11.4	0.06%
2009	19.3	0.10%
2010	31.4	0.15%
2011	60.6	0.27%

2012	96.7	0.43%
2013	134.5	0.58%
2014	185.9	0.79%
2015	253.0	1.05%

Early Days

The early development of solar technologies starting in the 1860s was driven by an expectation that coal would soon become scarce. Charles Fritts installed the world's first rooftop photovoltaic solar array, using 1%-efficient selenium cells, on a New York City roof in 1884. However, development of solar technologies stagnated in the early 20th century in the face of the increasing availability, economy, and utility of coal and petroleum. In 1974 it was estimated that only six private homes in all of North America were entirely heated or cooled by functional solar power systems. The 1973 oil embargo and 1979 energy crisis caused a reorganization of energy policies around the world and brought renewed attention to developing solar technologies. Deployment strategies focused on incentive programs such as the Federal Photovoltaic Utilization Program in the US and the Sunshine Program in Japan. Other efforts included the formation of research facilities in the United States (SERI, now NREL), Japan (NEDO), and Germany (Fraunhofer–ISE). Between 1970 and 1983 installations of photovoltaic systems grew rapidly, but falling oil prices in the early 1980s moderated the growth of photovoltaics from 1984 to 1996.

Mid-1990s to Early 2010s

In the mid-1990s, development of both, residential and commercial rooftop solar as well as utility-scale photovoltaic power stations, began to accelerate again due to supply issues with oil and natural gas, global warming concerns, and the improving economic position of PV relative to other energy technologies. In the early 2000s, the adoption of feed-in tariffs—a policy mechanism, that gives renewables priority on the grid and defines a fixed price for the generated electricity—lead to a high level of investment security and to a soaring number of PV deployments in Europe.

Current Status

For several years, worldwide growth of solar PV was driven by European deployment, but has since shifted to Asia, especially China and Japan, and to a growing number of countries and regions all over the world, including, but not limited to, Australia, Canada, Chile, India, Israel, Mexico, South Africa, South Korea, Thailand, and the United States.

Worldwide growth of photovoltaics has averaged 40% per year since 2000 and total installed capacity reached 139 GW at the end of 2013 with Germany having the most cumulative installations (35.7 GW) and Italy having the highest percentage of electricity generated by solar PV (7.0%). The largest manufacturers are located in China.

Concentrated solar power (CSP) also started to grow rapidly, increasing its capacity nearly tenfold from 2004 to 2013, albeit from a lower level and involving fewer countries than solar PV. As of the end of 2013, worldwide cumulative CSP-capacity reached 3,425 MW.

Forecasts

In 2010, the International Energy Agency predicted that global solar PV capacity could reach 3,000 GW or 11% of projected global electricity generation by 2050—enough to generate 4,500 TWh of electricity. Four years later, in 2014, the agency projected that, under its "high renewables" scenario, solar power could supply 27% of global electricity generation by 2050 (16% from PV and 11% from CSP). In 2015, analysts predicted that one million homes in the U.S. will have solar power by the end of 2016.

Photovoltaic Power Stations

The Desert Sunlight Solar Farm is a 550 MW power plant in Riverside County, California, that uses thin-film CdTe-modules made by First Solar. As of November 2014, the 550 megawatt Topaz Solar Farm was the largest photovoltaic power plant in the world. This was surpassed by the 579 MW Solar Star complex. The current largest photovoltaic power station in the world is Longyangxia Dam Solar Park, in Gonghe County, Qinghai, China.

World's largest photovoltaic power stations as of 2015

Name	Capacity (MW)	Location	Year Completed Info
Longyangxia Dam Solar Park	850	Qinghai, China	2013, 2015
Kamuthi Solar Power Project	648	Kamuthi, India	2015
Solar Star I and II	579	California, USA	2015
Topaz Solar Farm	550	California, USA	2014
Desert Sunlight Solar Farm	550	California, USA	2015
California Valley Solar Ranch	292	California, USA	2013
Agua Caliente Solar Project	290	Arizona, USA	2014
Mount Signal Solar	266	California, USA	2014
Antelope Valley Solar Ranch	266	California, USA	*pending*
Charanka Solar Park	224	Gujarat, India	2012
Mesquite Solar project	207	Arizona, USA	*pending (planned 700 MW)*
Huanghe Hydropower Golmud Solar Park	200	Qinghai, China	2011
Gonghe Industrial Park Phase I	200	Gonghe County, China	2013
Imperial Valley Solar Project	200	California, USA	2013

Concentrating Solar Power Stations

Commercial concentrating solar power (CSP) plants, also called "solar thermal power stations", were first developed in the 1980s. The 377 MW Ivanpah Solar Power Facility, located in California's Mojave Desert, is the world's largest solar thermal power plant project. Other large CSP plants include the Solnova Solar Power Station (150 MW), the Andasol solar power station (150 MW), and Extresol Solar Power Station (150 MW), all in Spain. The principal advantage of CSP is the ability to efficiently add thermal storage, allowing the dispatching of electricity over up to a 24-hour period. Since peak electricity demand typically occurs at about 5 pm, many CSP power plants use 3 to 5 hours of thermal storage.

Largest operational solar thermal power stations			
Name	**Capacity (MW)**	**Location**	**Notes**
Ivanpah Solar Power Facility	392	Mojave Desert, California, USA	Operational since February 2014. Located southwest of Las Vegas.
Solar Energy Generating Systems	354	Mojave Desert, California, USA	Commissioned between 1984 and 1991. Collection of 9 units.
Mojave Solar Project	280	Barstow, California, USA	Completed December 2014
Solana Generating Station	280	Gila Bend, Arizona, USA	Completed October 2013 Includes a 6th thermal energy storage
Genesis Solar Energy Project	250	Blythe, California, USA	Completed April 2014
Solaben Solar Power Station	200	Logrosán, Spain	Completed 2012–2013
Noor I	160	Morocco	Completed 2016
Solnova Solar Power Station	150	Seville, Spain	Completed in 2010
Andasol solar power station	150	Granada, Spain	Completed 2011. Includes a 7.5th thermal energy storage.
Extresol Solar Power Station	150	Torre de Miguel Sesmero, Spain	Completed 2010–2012 Extresol 3 includes a 7.5th thermal energy storage

Part of the 354 MW Solar Energy Generating Systems (SEGS) parabolic trough solar complex in northern San Bernardino County, California

Economics

Cost

Adjusting for inflation, it cost $96 per watt for a solar module in the mid-1970s. Process improvements and a very large boost in production have brought that figure down to 68 cents per watt in February 2016, according to data from Bloomberg New Energy Finance. Palo Alto California signed a wholesale purchase agreement in 2016 that secured solar power for 3.7 cents per kilowatt-hour. And in sunny Dubai large-scale solar generated electricity sold in 2016 for just 2.99

cents per kilowatt-hour -- "competitive with any form of fossil-based electricity — and cheaper than most."

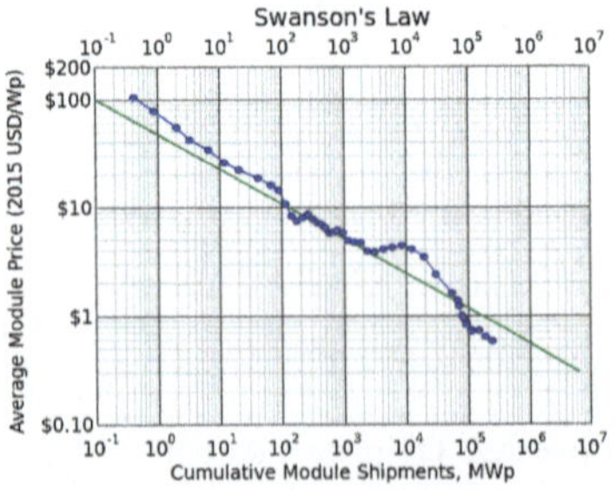

Swanson's law – the PV learning curve

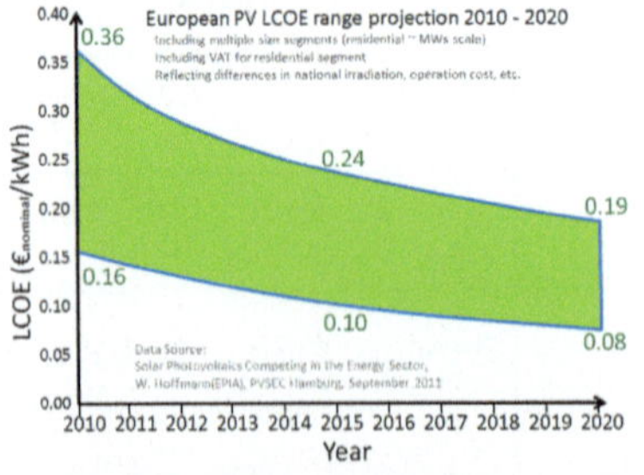

Solar PV – LCOE for Europe until 2020 (in euro-cts. per kWh)

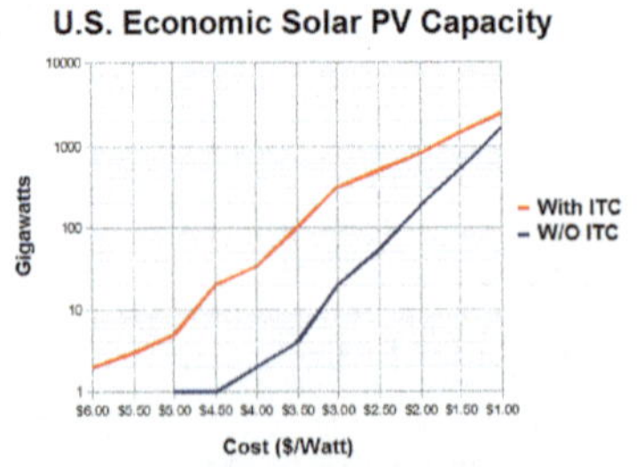

Economic photovoltaic capacity vs installation cost, in the United States

Photovoltaic systems use no fuel, and modules typically last 25 to 40 years. Thus, capital costs make up most of the cost of solar power. Operations and maintenance costs for new utility-scale solar plants in the US are estimated to be 9 percent of the cost of photovoltaic electricity, and 17 percent of the cost of solar thermal electricity. Governments have created various financial incentives to encourage the use of solar power, such as feed-in tariff programs. Also, Renewable portfolio standards impose a government mandate that utilities generate or acquire a certain percentage of renewable power regardless of increased energy procurement costs. In most states, RPS goals can be achieved by any combination of solar, wind, biomass, landfill gas, ocean, geothermal, municipal solid waste, hydroelectric, hydrogen, or fuel cell technologies.

Levelized Cost of Electricity

The PV industry is beginning to adopt levelized cost of electricity (LCOE) as the unit of cost. The electrical energy generated is sold in units of kilowatt-hours (kWh). As a rule of thumb, and depending on the local insolation, 1 watt-peak of installed solar PV capacity generates about 1 to 2 kWh of electricity per year. This corresponds to a capacity factor of around 10–20%. The product of the local cost of electricity and the insolation determines the break even point for solar power. The International Conference on Solar Photovoltaic Investments, organized by EPIA, has estimated that PV systems will pay back their investors in 8 to 12 years. As a result, since 2006 it has been economical for investors to install photovoltaics for free in return for a long term power purchase agreement. Fifty percent of commercial systems in the United States were installed in this manner in 2007 and over 90% by 2009.

Shi Zhengrong has said that, as of 2012, unsubsidised solar power is already competitive with fossil fuels in India, Hawaii, Italy and Spain. He said "We are at a tipping point. No longer are renewable power sources like solar and wind a luxury of the rich. They are now starting to compete in the real world without subsidies". "Solar power will be able to compete without subsidies against conventional power sources in half the world by 2015".

Current Installation Prices

In its 2014 edition of the *Technology Roadmap: Solar Photovoltaic Energy* report, the International Energy Agency (IEA) published prices for residential, commercial and utility-scale PV systems for eight major markets as of 2013. However, DOE's SunShot Initiative has reported much lower U.S. installation prices. In 2014, prices continued to decline. The SunShot Initiative modeled U.S. system prices to be in the range of $1.80 to $3.29 per watt. Other sources identify similar price ranges of $1.70 to $3.50 for the different market segments in the U.S., and in the highly penetrated German market, prices for residential and small commercial rooftop systems of up to 100 kW declined to $1.36 per watt (€1.24/W) by the end of 2014. In 2015, Deutsche Bank estimated costs for small residential rooftop systems in the U.S. around $2.90 per watt. Costs for utility-scale systems in China and India were estimated as low as $1.00 per watt.

Typical PV system prices in 2013 in selected countries (USD)								
USD/W	Australia	China	France	Germany	Italy	Japan	United Kingdom	United States
Residential	1.8	1.5	4.1	2.4	2.8	4.2	2.8	4.9[1]
Commercial	1.7	1.4	2.7	1.8	1.9	3.6	2.4	4.5[1]
Utility-scale	2.0	1.4	2.2	1.4	1.5	2.9	1.9	3.3[1]

Grid Parity

Grid parity, the point at which the cost of photovoltaic electricity is equal to or cheaper than the price of grid power, is more easily achieved in areas with abundant sun and high costs for electricity such as in California and Japan. In 2008, The levelized cost of electricity for solar PV was $0.25/kWh or less in most of the OECD countries. By late 2011, the fully loaded cost was predicted to fall below $0.15/kWh for most of the OECD and to reach $0.10/kWh in sunnier regions. These cost levels are driving three emerging trends: vertical integration of the supply chain, origination of power purchase agreements (PPAs) by solar power companies, and unexpected risk for traditional power generation companies, grid operators and wind turbine manufacturers.

Grid parity was first reached in Spain in 2013, Hawaii and other islands that otherwise use fossil fuel (diesel fuel) to produce electricity, and most of the US is expected to reach grid parity by 2015.

In 2007, General Electric's Chief Engineer predicted grid parity without subsidies in sunny parts of the United States by around 2015; other companies predicted an earlier date: the cost of solar power will be below grid parity for more than half of residential customers and 10% of commercial customers in the OECD, as long as grid electricity prices do not decrease through 2010.

Productivity by Location

Different parts of the world experience different amounts of sunshine, depending on latitude and weather. Locations nearer the equator receive many more hours of sunshine than those further north or south, thus photovoltaic panels produce more kWhs per year, and so can be far more economically desirable in some places more than others.

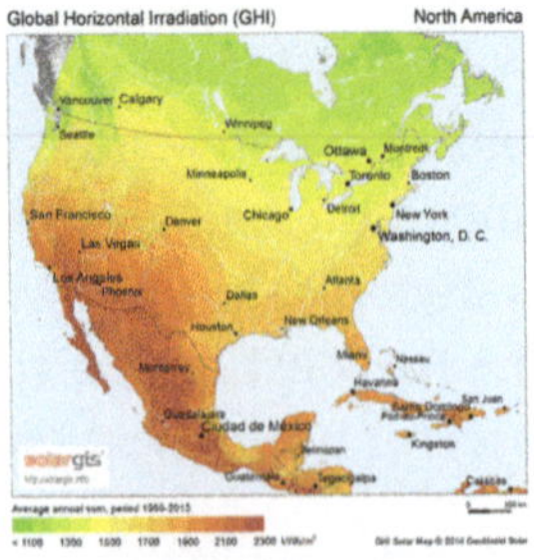

North America

South America

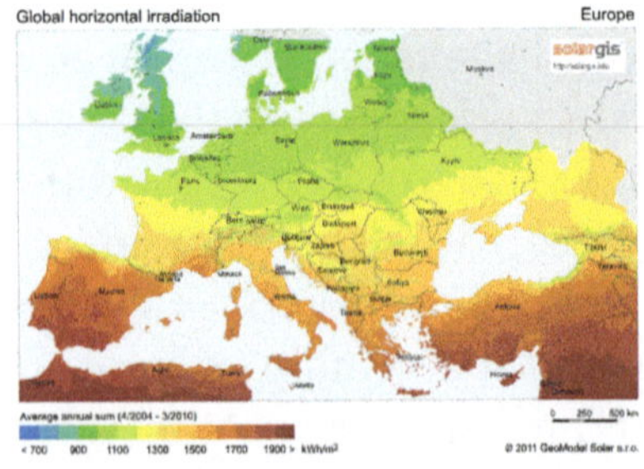

Europe

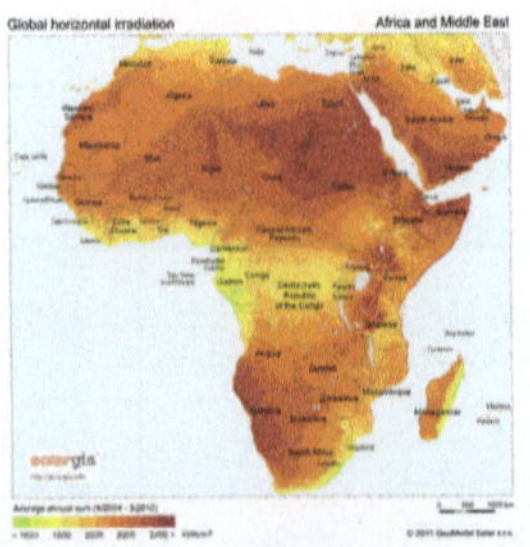

Africa and Middle East

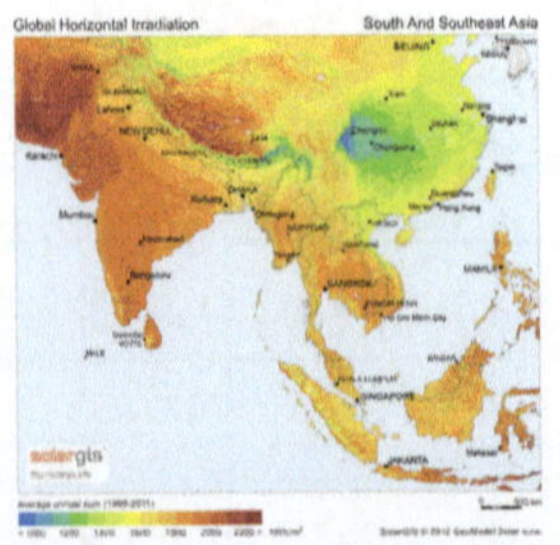

South and South-East Asia

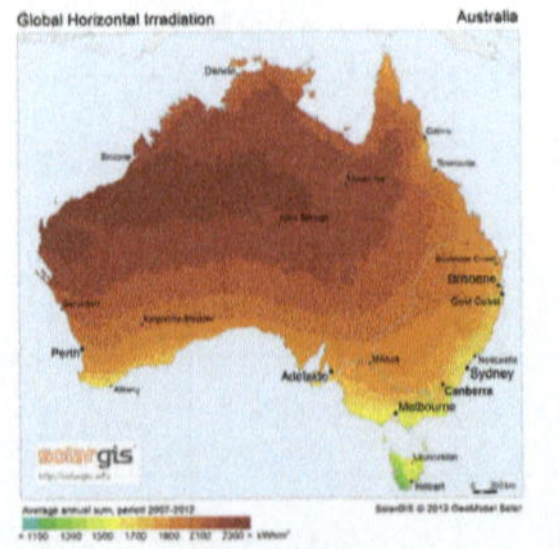

Australia

Self Consumption

In cases of self consumption of the solar energy, the payback time is calculated based on how much electricity is not purchased from the grid. For example, in Germany, with electricity prices of 0.25 €/kWh and insolation of 900 kWh/kW, one kWp will save €225 per year, and with an installation cost of 1700 €/KWp the system cost will be returned in less than seven years. However, in many cases, the patterns of generation and consumption do not coincide, and some or all of the energy is fed back into the grid. The electricity is sold, and at other times when energy is taken from the grid, electricity is bought. The relative costs and prices obtained affect the economics. In many markets, the price paid for sold PV electricity is significantly lower than the price of bought electricity, which incentivizes self consumption. Moreover, separate self consumption incentives have been used in e.g. Germany and Italy. Grid interaction regulation has also included limitations of grid feed-in in some regions in Germany with high amounts of installed PV capacity. By increasing self consumption, the grid feed-in can be limited without curtailment, which wastes electricity.

A good match between generation and consumption is key for high self consumption, and should be considered when deciding where to install solar power and how to dimension the installation. The match can be improved with batteries or controllable electricity consumption. However, batteries are expensive and profitability may require provision of other services from them besides self consumption increase. Hot water storage tanks with electric heating with heat pumps or resistance heaters can provide low-cost storage for self consumption of solar power. Shiftable loads, such as dishwashers, tumble dryers and washing machines, can provide controllable consumption with only a limited effect on the users, but their effect on self consumption of solar power may be limited.

Energy Pricing and Incentives

The political purpose of incentive policies for PV is to facilitate an initial small-scale deployment to begin to grow the industry, even where the cost of PV is significantly above grid parity, to allow the industry to achieve the economies of scale necessary to reach grid parity. The policies are implemented to promote national energy independence, high tech job creation and reduction of CO_2 emissions. Three incentive mechanisms are often used in combination as investment subsidies: the authorities refund part of the cost of installation of the system, the electricity utility buys PV electricity from the producer under a multiyear contract at a guaranteed rate, and Solar Renewable Energy Certificates (SRECs).

Rebates

With investment subsidies, the financial burden falls upon the taxpayer, while with feed-in tariffs the extra cost is distributed across the utilities' customer bases. While the investment subsidy may be simpler to administer, the main argument in favour of feed-in tariffs is the encouragement of quality. Investment subsidies are paid out as a function of the nameplate capacity of the installed system and are independent of its actual power yield over time, thus rewarding the overstatement of power and tolerating poor durability and maintenance. Some electric companies offer rebates to their customers, such as Austin Energy in Texas, which offers $2.50/watt installed up to $15,000.

Net Metering

In net metering the price of the electricity produced is the same as the price supplied to the consumer, and the consumer is billed on the difference between production and consumption. Net metering can usually be done with no changes to standard electricity meters, which accurately measure power in both directions and automatically report the difference, and because it allows homeowners and businesses to generate electricity at a different time from consumption, effectively using the grid as a giant storage battery. With net metering, deficits are billed each month while surpluses are rolled over to the following month. Best practices call for perpetual roll over of kWh credits. Excess credits upon termination of service are either lost, or paid for at a rate ranging from wholesale to retail rate or above, as can be excess annual credits. In New Jersey, annual excess credits are paid at the wholesale rate, as are left over credits when a customer terminates service.

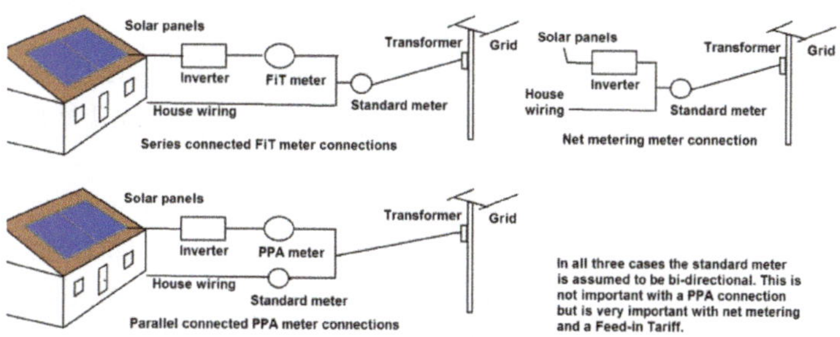

Understanding Feed-in Tariff and Power Purchase Agreement meter connections

Net metering, unlike a feed-in tariff, requires only one meter, but it must be bi-directional.

Feed-in Tariffs (FIT)

With feed-in tariffs, the financial burden falls upon the consumer. They reward the number of kilowatt-hours produced over a long period of time, but because the rate is set by the authorities, it may result in perceived overpayment. The price paid per kilowatt-hour under a feed-in tariff exceeds the price of grid electricity. Net metering refers to the case where the price paid by the utility is the same as the price charged.

The complexity of approvals in California, Spain and Italy has prevented comparable growth to Germany even though the return on investment is better. In some countries, additional incentives are offered for BIPV compared to stand alone PV.

- France + EUR 0.16 /kWh (compared to semi-integrated) or + EUR 0.27/kWh (compared to stand alone)
- Italy + EUR 0.04-0.09 kWh
- Germany + EUR 0.05/kWh (facades only)

Solar Renewable Energy Credits (SRECs)

Alternatively, SRECs allow for a market mechanism to set the price of the solar generated electricity subsity. In this mechanism, a renewable energy production or consumption target is set, and the utility (more technically the Load Serving Entity) is obliged to purchase renewable energy or face a fine (Alternative Compliance Payment or ACP). The producer is credited for an SREC for every 1,000 kWh of electricity produced. If the utility buys this SREC and retires it, they avoid paying the ACP. In principle this system delivers the cheapest renewable energy, since the all solar facilities are eligible and can be installed in the most economic locations. Uncertainties about the future value of SRECs have led to long-term SREC contract markets to give clarity to their prices and allow solar developers to pre-sell and hedge their credits.

Financial incentives for photovoltaics differ across countries, including Australia, China, Germany, Israel, Japan, and the United States and even across states within the US.

The Japanese government through its Ministry of International Trade and Industry ran a successful programme of subsidies from 1994 to 2003. By the end of 2004, Japan led the world in installed PV capacity with over 1.1 GW.

In 2004, the German government introduced the first large-scale feed-in tariff system, under the German Renewable Energy Act, which resulted in explosive growth of PV installations in Germany. At the outset the FIT was over 3x the retail price or 8x the industrial price. The principle behind the German system is a 20-year flat rate contract. The value of new contracts is programmed to decrease each year, in order to encourage the industry to pass on lower costs to the end users. The programme has been more successful than expected with over 1GW installed in 2006, and political pressure is mounting to decrease the tariff to lessen the future burden on consumers.

Subsequently, Spain, Italy, Greece—that enjoyed an early success with domestic solar-thermal installations for hot water needs—and France introduced feed-in tariffs. None have replicated the programmed decrease of FIT in new contracts though, making the German incentive relatively less and less attractive compared to other countries. The French and Greek FIT offer a high premium (EUR 0.55/kWh) for building integrated systems. California, Greece, France and Italy have 30-50% more insolation than Germany making them financially more attractive. The

Greek domestic "solar roof" programme (adopted in June 2009 for installations up to 10 kW) has internal rates of return of 10-15% at current commercial installation costs, which, furthermore, is tax free.

In 2006 California approved the 'California Solar Initiative', offering a choice of investment subsidies or FIT for small and medium systems and a FIT for large systems. The small-system FIT of $0.39 per kWh (far less than EU countries) expires in just 5 years, and the alternate "EPBB" residential investment incentive is modest, averaging perhaps 20% of cost. All California incentives are scheduled to decrease in the future depending as a function of the amount of PV capacity installed.

At the end of 2006, the Ontario Power Authority (OPA, Canada) began its Standard Offer Program, a precursor to the Green Energy Act, and the first in North America for distributed renewable projects of less than 10 MW. The feed-in tariff guaranteed a fixed price of $0.42 CDN per kWh over a period of twenty years. Unlike net metering, all the electricity produced was sold to the OPA at the given rate.

Grid Integration

Construction of the Salt Tanks which provide efficient thermal energy storage so that output can be provided after the sun goes down, and output can be scheduled to meet demand requirements. The 280 MW Solana Generating Station is designed to provide six hours of energy storage. This allows the plant to generate about 38 percent of its rated capacity over the course of a year.

Pumped-storage hydroelectricity (PSH). This facility in Geesthacht, Germany, also includes a solar array.

The overwhelming majority of electricity produced worldwide is used immediately, since storage is usually more expensive and because traditional generators can adapt to demand. However both solar power and wind power are variable renewable energy, meaning that all available output must be taken when it is available, and either stored for *when it can be used later*, or transported over transmission lines to *where it can be used now*. Since solar energy is not available at night, storing its energy is potentially an important issue particularly in off-grid and for future 100% renewable energy scenarios to have continuous energy availability. In addition, integrating large amounts of solar power with existing generation equipment has caused issues in some cases. For example, in Germany, California and Hawaii, electricity prices have been known to go negative when solar is generating a lot of power causing interference with existing baseload generation contracts.

Concentrated solar power plants may use thermal energy storage to store the solar energy, such as in high-temperature molten salts. These salts are an effective storage medium because they are low-cost, have a high specific heat capacity, and can deliver heat at temperatures compatible with conventional power systems. This method of energy storage is used, for example, by the Solar Two power station, allowing it to store 1.44 TJ in its 68 m^3 storage tank, enough to provide full output for close to 39 hours, with an efficiency of about 99%.

Rechargeable batteries have been traditionally used to store excess electricity in stand alone PV systems. With grid-connected photovoltaic power system, excess electricity can be sent to the electrical grid. Net metering and feed-in tariff programs give these systems a credit for the electricity they produce. This credit offsets electricity provided from the grid when the system cannot meet demand, effectively trading with the grid instead of storing excess electricity. Credits are normally rolled over from month to month and any remaining surplus settled annually. When wind and solar are a small fraction of the grid power, other generation techniques can adjust their output appropriately, but as these forms of variable power grow, additional balance on the grid is needed. As prices are rapidly declining, PV systems increasingly use rechargeable batteries to store a surplus to be later used at night. Batteries used for grid-storage stabilize the electrical grid by leveling out peak loads usually for several minutes, and in rare cases for hours. In the future, less expensive batteries could play an important role on the electrical grid, as they can charge during periods when generation exceeds demand and feed their stored energy into the grid when demand is higher than generation.

Although not permitted under the US National Electric Code, it is technically possible to have a "plug and play" PV microinverter. A careful system design would enable such systems to meet all technical, though not all safety requirements. There are several companies selling plug and play solar systems available on the web, but there is a concern that if people install their own it will reduce the enormous labor advantage solar has over fossil fuels.

Common battery technologies used in today's home PV systems include, the valve regulated lead-acid battery– a modified version of the conventional lead–acid battery, nickel–cadmium and lithium-ion batteries. Lead-acid batteries are currently the predominant technology used in small-scale, residential PV systems, due to their high reliability, low self discharge and investment and maintenance costs, despite shorter lifetime and lower energy density. However, lithium-ion batteries have the potential to replace lead-acid batteries in the near future, as they are being intensively developed and lower prices are expected due to economies of scale provided by large production facilities such as the Gigafactory 1. In addition, the Li-ion batteries of plug-in electric cars may serve as a future storage devices in a vehicle-to-grid system. Since most vehicles are parked an average of 95 percent of the time, their batteries could be used to let electricity flow from the car to the power lines and back. Other rechargeable batteries used for distributed PV systems include, sodium–sulfur and vanadium redox batteries, two prominent types of a molten salt and a flow battery, respectively.

Conventional hydroelectricity works very well in conjunction with variable electricity sources such as solar and wind, the water can be held back and allowed to flow as required. Where a suitable river is not available, pumped-storage hydroelectricity stores energy in the form of water pumped when surplus electricity is available, from a lower elevation reservoir to a higher elevation one. The energy is recovered when demand is high by releasing the water through a hydroelectric power generator. However, this cycle can lose 20% of the energy to round trip inefficiencies.

The combination of wind and solar PV has the advantage that the two sources complement each other because the peak operating times for each system occur at different times of the day and year. The power generation of such solar hybrid power systems is therefore more constant and fluctuates less than each of the two component subsystems. Solar power is seasonal, particularly in northern/southern climates, away from the equator, suggesting a need for long term seasonal storage in a medium such as hydrogen or pumped hydroelectric. The Institute for Solar Energy Supply Technology of the University of Kassel pilot-tested a combined power plant linking solar, wind, biogas and hydrostorage to provide load-following power from renewable sources.

Research is also undertaken in this field of artificial photosynthesis. It involves the use of nanotechnology to store solar electromagnetic energy in chemical bonds, by splitting water to produce hydrogen fuel or then combining with carbon dioxide to make biopolymers such as methanol. Many large national and regional research projects on artificial photosynthesis are now trying to develop techniques integrating improved light capture, quantum coherence methods of electron transfer and cheap catalytic materials that operate under a variety of atmospheric conditions. Senior researchers in the field have made the public policy case for a Global Project on Artificial Photosynthesis to address critical energy security and environmental sustainability issues.

Environmental Impacts

Part of the Senftenberg Solarpark, a solar photovoltaic power plant located on former open-pit mining areas close to the city of Senftenberg, in Eastern Germany. The 78 MW Phase 1 of the plant was completed within three months.

Unlike fossil fuel based technologies, solar power does not lead to any harmful emissions during operation, but the production of the panels leads to some amount of pollution.

Greenhouse Gases

The life-cycle greenhouse-gas emissions of solar power are in the range of 22 to 46 gram (g) per kilowatt-hour (kWh) depending on if solar thermal or solar PV is being analyzed, respectively. With this potentially being decreased to 15 g/kWh in the future. For comparison (of weighted averages), a combined cycle gas-fired power plant emits some 400–599 g/kWh, an oil-fired power plant 893 g/kWh, a coal-fired power plant 915–994 g/kWh or with carbon capture and storage some 200 g/kWh, and a geothermal high-temp. power plant 91–122 g/kWh. The life cycle emission intensity of hydro, wind and nuclear power are lower than solar's. Similar to all energy sources were their total life cycle emissions primarily lay in the construction and transportation phase, the switch to low carbon power in the manufacturing and transportation of solar devices would further reduce carbon emissions. BP Solar owns two factories built by Solarex (one in Maryland, the other in Virginia) in which all of the energy used to manufacture solar panels is produced by solar panels. A 1-kilowatt system eliminates the burning of approximately 170 pounds of coal, 300 pounds of carbon dioxide from being released into the atmosphere, and saves up to 105 gallons of water consumption monthly.

The US National Renewable Energy Laboratory (NREL), in harmonizing the disparate estimates of life-cycle GHG emissions for solar PV, found that the most critical parameter was the solar insolation of the site: GHG emissions factors for PV solar are inversely proportional to insolation. For a site with insolation of 1700 kWh/m2/year, typical of southern Europe, NREL researchers estimated GHG emissions of 45 gCO_2e/kWh. Using the same assumptions, at Phoenix, USA, with insolation of 2400 kWh/m2/year, the GHG emissions factor would be reduced to 32 g of CO_2e/kWh.

The New Zealand Parliamentary Commissioner for the Environment found that the solar PV would have little impact on the country's greenhouse gas emissions. The country already generates 80 percent of its electricity from renewable resources (primarily hydroelectricity and geothermal) and national electricity usage peaks on winter evenings whereas solar generation peaks on summer afternoons, meaning a large uptake of solar PV would end up displacing other renewable generators before fossil-fueled power plants.

Energy Payback

The energy payback time (EPBT) of a power generating system is the time required to generate as much energy as is consumed during production and lifetime operation of the system. Due to improving production technologies the payback time has been decreasing constantly since the introduction of PV systems in the energy market. In 2000 the energy payback time of PV systems was estimated as 8 to 11 years and in 2006 this was estimated to be 1.5 to 3.5 years for crystalline silicon silicon PV systems and 1–1.5 years for thin film technologies (S. Europe). These figures fell to 0.75–3.5 years in 2013, with an average of about 2 years for crystalline silicon PV and CIS systems.

Another economic measure, closely related to the energy payback time, is the energy returned on energy invested (EROEI) or energy return on investment (EROI), which is the ratio of electricity generated divided by the energy required to build *and maintain* the equipment. (This is not the same as the economic return on investment (ROI), which varies according to local energy prices, subsidies available and metering techniques.) With expected lifetimes of 30 years, the EROEI of PV systems are in the range of 10 to 30, thus generating enough energy over their lifetimes to reproduce themselves many times (6-31 reproductions) depending on what type of material, balance of system (BOS), and the geographic location of the system.

Other Issues

One issue that has often raised concerns is the use of cadmium (Cd), a toxic heavy metal that has the tendency to accumulate in ecological food chains. It is used as semiconductor component in CdTe solar cells and as buffer layer for certain CIGS cells in the form of CdS. The amount of cadmium used in thin-film PV modules is relatively small (5–10 g/m^2) and with proper recycling and emission control techniques in place the cadmium emissions from module production can be almost zero. Current PV technologies lead to cadmium emissions of 0.3–0.9 microgram/kWh over the whole life-cycle. Most of these emissions actually arise through the use of coal power for the manufacturing of the modules, and coal and lignite combustion leads to much higher emissions of cadmium. Life-cycle cadmium emissions from coal is 3.1 microgram/kWh, lignite 6.2, and natural gas 0.2 microgram/kWh.

In a life-cycle analysis it has been noted, that if electricity produced by photovoltaic panels were used to manufacture the modules instead of electricity from burning coal, cadmium emissions from coal power usage in the manufacturing process could be entirely eliminated.

In the case of crystalline silicon modules, the solder material, that joins together the copper strings of the cells, contains about 36 percent of lead (Pb). Moreover, the paste used for screen printing front and back contacts contains traces of Pb and sometimes Cd as well. It is estimated that about 1,000 metric tonnes of Pb have been used for 100 gigawatts of c-Si solar modules. However, there is no fundamental need for lead in the solder alloy.

Some media sources have reported that concentrated solar power plants have injured or killed large numbers of birds due to intense heat from the concentrated sunrays. This adverse effect does not apply to PV solar power plants, and some of the claims may have been overstated or exaggerated.

A 2014-published life-cycle analysis of land use for various sources of electricity concluded that the large-scale implementation of solar and wind potentially reduces pollution-related environmental impacts. The study found that the land-use footprint, given in square meter-years per megawatt-hour (m^2a/MWh), was lowest for wind, natural gas and rooftop PV, with 0.26, 0.49 and 0.59, respectively, and followed by utility-scale solar PV with 7.9. For CSP, the footprint was 9 and 14, using parabolic troughs and solar towers, respectively. The largest footprint had coal-fired power plants with 18 m^2a/MWh.

Emerging Technologies

Concentrator Photovoltaics

Concentrator photovoltaics (CPV) systems employ sunlight concentrated onto photovoltaic surfaces for the purpose of electrical power production. Contrary to conventional photovoltaic systems, it uses lenses and curved mirrors to focus sunlight onto small, but highly efficient, multi-junction solar cells. Solar concentrators of all varieties may be used, and these are often mounted on a solar tracker in order to keep the focal point upon the cell as the sun moves across the sky. Luminescent solar concentrators (when combined with a PV-solar cell) can also be regarded as a CPV system. Concentrated photovoltaics are useful as they can improve efficiency of PV-solar panels drastically.

CPV modules on dual axis solar trackers in Golmud, China

In addition, most solar panels on spacecraft are also made of high efficient multi-junction photovoltaic cells to derive electricity from sunlight when operating in the inner Solar System.

Floatovoltaics

Floatovoltaics are an emerging form of PV systems that float on the surface of irrigation canals, water reservoirs, quarry lakes, and tailing ponds. Several systems exist in France, India, Japan, Korea, the United Kingdom and the United States. These systems reduce the need of valuable land area, save drinking water that would otherwise be lost through evaporation, and show a higher efficiency of solar energy conversion, as the panels are kept at a cooler temperature than they would be on land. Although not floating, other dual-use facilities with solar power include fisheries.

Active Solar

Solar trackers may be driven by active or passive solar technology

Solar hot water systems use pumps or fans to circulate fluid (often a mixture of water and glycol to prevent freezing during winter periods) or air, through solar collectors, and are therefore classified under active solar technology.

Some of the basic benefits of active systems is that controls (usually electrical) can be used to maximize their effectiveness. For example, a passive solar thermal array which does not rely on pumps and sensors will only start circulating when a certain amount of internal energy has built up in the system. Using sensors and pumps, a relatively small amount of energy (i.e. that used to power a pump and controller) can harvest a far larger amount of available thermal energy by switching on as soon as a useful temperature differential becomes present. Controls also allow a greater variety of choices for utilizing the energy that becomes available. For example, a solar thermal array could heat a swimming pool on a relatively cool morning where heating a domestic hot water cylinder was impractical due to the different stored water temperatures. Later in the day as the temperature rises the controls could be used to switch the solar heated water over to the cylinder instead.

Disadvantages of active solar systems include that the external power sources can fail (probably rendering them useless) and that controls require maintenance.

Most solar collectors are fixed in their array position mounting, but can have a higher performance if they track the path of the sun through the sky (however it is unusual for thermal collectors to be mounted in this way). Solar trackers, used to orient solar arrays may be driven by either passive or active technology, and can have a significant gain in energy yield over the course of a year when compared to a fixed array. Again passive solar tracking would rely on the inherent thermo-dynamic properties of the materials used in the system rather than an external power source to generate its tracking movement. Active Solar Tracking would utilise sensors and motors track the path of the sun across the sky. This action can be caused by geographical and time data being programmed into the controls. However, some systems actually track the brightest point in the sky using light sensors, and manufacturers claim this can add a significant extra yield over and above geographical tracking.

References

- Salpakari, Jyri; Lund, Peter (2016). "Optimal and rule-based control strategies for energy flexibility in buildings with PV". Applied Energy. 161: 425–436. doi:10.1016/j.apenergy.2015.10.036
- Anderson, Lorraine; Palkovic, Rick (1994). Cooking with Sunshine (The Complete Guide to Solar Cuisine with 150 Easy Sun-Cooked Recipes). Marlowe & Company. ISBN 1-56924-300-X
- Leon, M.; Kumar, S. (2007). "Mathematical modeling and thermal performance analysis of unglazed transpired solar collectors". Solar Energy. 81 (1): 62–75. Bibcode:2007SoEn...81...62L. doi:10.1016/j.solener.2006.06.017
- International Energy Agency (2014). "Technology Roadmap: Solar Photovoltaic Energy" (PDF). IEA. Archived (PDF) from the original on 7 October 2014. Retrieved 7 October 2014
- Martin, Christopher L.; Goswami, D. Yogi (2005). Solar Energy Pocket Reference. International Solar Energy Society. ISBN 0-9771282-0-2
- Tabor, H. Z.; Doron, B. (1990). "The Beith Ha'Arava 5 MW(e) Solar Pond Power Plant (SPPP)--Progress Report". Solar Energy. 45 (4): 247–253. Bibcode:1990SoEn...45..247T. doi:10.1016/0038-092X(90)90093-R
- Karan, Kaul; et. al. (2001). "Row Orientation Affects Fruit Yield in Field-Grown Okra". Journal of Sustainable Agriculture. 17 (2/3): 169–174. doi:10.1300/J064v17n02_14

- "An Assessment of Solar Energy Conversion Technologies and Research Opportunities" (PDF). Stanford University – Global Climate Change & Energy Project. Retrieved 2 July 2008
- Smil, Vaclav (2006). Energy at the Crossroads (PDF). Organisation for Economic Co-operation and Development. ISBN 0-262-19492-9
- "Electric cars not solar panels, says Environment Commissioner". Parliamentary Commissioner for the Environment. 22 March 2016. Retrieved 23 March 2016
- Zedtwitz, P.V.; et. al. (2006). "Hydrogen production via the solar thermal decarbonization of fossil fuels". Solar Energy. 80 (10): 1333–7. Bibcode:2006SoEn...80.1333Z. doi:10.1016/j.solener.2005.06.007

4

Biomass Energy: An Overview

Biomass is energy that is generated by burning organic matter. Some of the methods used to convert biomass are chemical, biochemical and thermal conversions. Anaerobic digestion, biofuel and biogas are some of the topics discussed. This chapter is an overview of the subject matter incorporating all the major aspects of bioenergy.

Biomass Energy

In the past few years, there have been significant improvements in renewable energy technologies along with declines in cost. The growing concern for the environment and sustainable development, have led to worldwide interest in renewable energies and bio- energy in particular. Biomass can be converted into modern energy forms such as liquid and gaseous fuels, electricity, and process heat to provide energy services needed by rural and urban populations and also by industry. This paper explains the different ways of extracting energy from biomass and a comparison is made among them. This paper also explains about the potentiality of biomass energy in India, analyses current situation compares bio-energy and other options for promoting development, brings out the advantages over the other renewables putting forth the drawbacks to be overcome to make it still more successful. This paper analyses current situation compares bio-energy and other options for promoting development, explore the potential for bio-energy.

Introduction

In past 10 years or so, considerable practical experience has accumulated in India as well as in other developing and industrialized countries, on biomass energy production and conversion. India is pioneer among developing countries, with significant indigenous efforts in promoting renewable energy technologies. The importance of bio- energy as a modern fuel has been recognized. India has about 70,000 villages yet to be connected to the electricity grid. The supply of grid power to rural areas is characterized by

a) Low loads

b) Power shortages

c) Low reliability

d) Low and fluctuating voltages

e) High transmission and distribution costs and power losses

Decentralized power generation based on renewables is an attractive option to meet the energy needs. The availability of biomass such as wood, cow-dung, leaf litter in rural areas is more. Hence a choice of biomass energy especially in rural areas is more reasonable but at the same time the technology is being developed to meet the large- scale requirements using biomass.

Biomass energy has played a key role in the time of Second World War when there was a fuel deficiency. Many vehicles, tractors and trucks used wood gasifies, which generate producer's gas, running an internal combustion Engine. One of the major advantages of biomass energy is that it can be used in different forms. For e.g., Gas generated from the biomass can be directly used for cooking or it can be used for running an internal combustion Engine for developing stationary shaft power or otherwise coupled to generator for generating electric power.

The subsequent sections explain about the different ways of extracting energy from biomass, explaining about technological and economic aspects followed by a case study. The issue of land availability for biomass (wood) production.

Motivation

There are several renewable and non-renewable energy options for power generation at the decentralized level. It is necessary to understand why biomass-based energy options should receive priority over other options and to discuss the advantages to local and global communities as well as the environment.

Biomass is renewable fuel used in nearly every corner of the developing countries as a source of heat, particularly in the domestic sector. Biomass energy includes energy from all plant matter (tree, shrub, and crop) and animal dung. Biomass, unlike other renewables, is a versatile source of energy, which can be converted to modern forms such as liquid and gaseous fuels, electricity, and process heat. For example, small-scale (5-100 KW), medium-scale (1-10 MW), and large scale (about 50 MW).

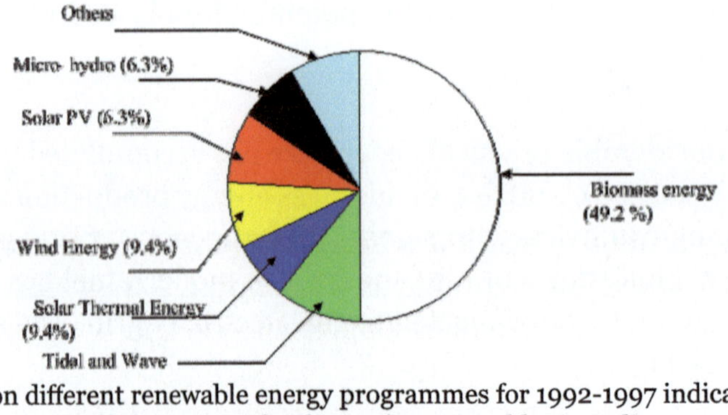

Expenditure on different renewable energy programmes for 1992-1997 indicating the share of bioenergy technologies for renewables in India

It can be seen from the figure the importance of biomass energy in Indian context.

Advantages of biomass gasify energy over other renewable energy options:

1. Suitable in most locations
2. Varying capacity can be installed; any capacity can be operated, even at lower loads; no seasonality.
3. Need for storage of energy is not required.

Advantages and drawbacks of biogas energy over other renewable energy options:

1. It can be used directly for cooking, or heating water from the abundantly available dung and dried plant leaves in rural areas.
2. Capacity determined by availability of dung. Not suitable for varying loads.
3. Not feasible to locate at all the locations.

Different Ways of Extracting Energy from Biomass

The different methods of biomass extraction can be broadly be classified as:

1. Anaerobic Digestion
2. Gasification
3. Liquefaction

(a) Solid fuel combustion

The simplest and most common way of extracting energy from biomass is by direct combustion of solid matter. Majority of the developing countries especially in rural areas obtain the majority of their energy needs from the burning of wood, animal dung and other biomass. But burning can be inefficient. An open fireplace may let large amounts of heat escape, while a significant proportion of the fuel may not even get burnt.

(b) Gasification

Gasification is a process that exposes a solid fuel to high temperatures and limited oxygen, to produce a gaseous fuel. This is a mix of gases such as carbon monoxide, carbon dioxide, nitrogen, hydrogen and methane.

Gasification has several advantages over burning solid fuel. One is convenience – one of the resultant gases, methane, can be treated in a similar way as natural gas, and used for the same purposes.

Another advantage of gasification is that it produces a fuel that has had many impurities removed and will therefore cause fewer pollution problems when burnt. And, under suitable circumstances, it can produce synthesis gas, a mixture of carbon monoxide and hydrogen. This can be used to make almost any hydrocarbon (e.g., methane and methanol), which can then be substituted for fossil fuels. But hydrogen itself is a potential fuel of the future.

(c) Paralysis

Paralysis is an old technology with a new lease of life. In its simplest form it involves heating the biomass to drive off the volatile matter, leaving behind the black residue we know as charcoal. This has double the energy density of the original material. This means that charcoal, which is half the weight of the original biomass, contains the same amount of energy – making the fuel more transportable. The charcoal also burns at a much higher temperature than the original biomass, making it more useful for manufacturing processes. More sophisticated Paralysis techniques have been developed recently to collect the volatiles that are otherwise lost to the system. The collected volatiles produce a gas rich in hydrogen (a potential fuel) and carbon monoxide. These compounds, if desired, can be synthesized into methane, methanol and other hydrocarbons. 'Flash' Paralysis can be used to produce bio-crude – a combustible fuel.

(d) Digestion

Biomass digestion works by the action of anaerobic bacteria. These microorganisms usually live at the bottom of swamps or in other places where there is no air, consuming dead organic matter to produce, among other things, methane and hydrogen.

We can put these bacteria to work for us. By feeding organic matter such as animal dung or human sewage into tanks – called digesters - and adding bacteria, we can collect the emitted gas to use as an energy source. This can be a very efficient means of extracting usable energy from such biomass – up to two-thirds of the fuel energy of the animal dung is recovered.

Another, related, technique is to collect gas from landfill sites. A large proportion of household biomass waste, such as kitchen scraps, lawn clippings and pruning, ends up at the local tip. Over a period of several decades, anaerobic bacteria are at work at the bottom of such tips, steadily decomposing the organic matter and emitting methane. The gas can be extracted and used by 'capping' a landfill site with an impervious layer of clay and then inserting perforated pipes that collect the gas and bring it to the surface.

(e) Fermentation

Like many of the other processes described here, fermentation isn't a new idea. For centuries, people have used yeasts and other microorganisms to ferment the sugar of various plants into ethanol. Producing fuel from biomass by fermentation is just an extension of this old process, although a wider range of plant material can now be used, from sugar cane to wood fiber. For instance, the waste from a wheat mill in New South Wales has been used to produce ethanol through fermentation. This is then mixed with diesel to produce 'dishelm', a product used by some trucks and buses in Sydney and Canberra.

An elaborated discussion on Digestion and Gasification, which are the major ways employed in India, are explained in subsequent sections.

Anaerobic Digestion

Anaerobic digestion is a collection of processes by which microorganisms break down biodegradable material in the absence of oxygen. The process is used for industrial or domestic purposes to manage waste or to produce fuels. Much of the fermentation used industrially to produce food and drink products, as well as home fermentation, uses anaerobic digestion.

Anaerobic digestion occurs naturally in some soils and in lake and oceanic basin sediments, where it is usually referred to as "anaerobic activity". This is the source of marsh gas methane as discovered by Volta in 1776.

The digestion process begins with bacterial hydrolysis of the input materials. Insoluble organic polymers, such as carbohydrates, are broken down to soluble derivatives that become available for other bacteria. Acidogenic bacteria then convert the sugars and amino acids into carbon dioxide, hydrogen, ammonia, and organic acids. These bacteria convert these resulting organic acids into acetic acid, along with additional ammonia, hydrogen, and carbon dioxide. Finally, methanogens convert these products to methane and carbon dioxide. The methanogenic archaea populations play an indispensable role in anaerobic wastewater treatments.

Anaerobic digestion is used as part of the process to treat biodegradable waste and sewage sludge. As part of an integrated waste management system, anaerobic digestion reduces the emission of landfill gas into the atmosphere. Anaerobic digesters can also be fed with purpose-grown energy crops, such as maize.

Anaerobic digestion is widely used as a source of renewable energy. The process produces a biogas, consisting of methane, carbon dioxide and traces of other 'contaminant' gases. This biogas can be used directly as fuel, in combined heat and power gas engines or upgraded to natural gas-quality biomethane. The nutrient-rich digestate also produced can be used as fertilizer.

With the re-use of waste as a resource and new technological approaches that have lowered capital costs, anaerobic digestion has in recent years received increased attention among governments in a number of countries, among these the United Kingdom (2011), Germany and Denmark (2011).

Process

Many microorganisms affect anaerobic digestion, including acetic acid-forming bacteria (acetogens) and methane-forming archaea (methanogens). These organisms promote a number of chemical processes in converting the biomass to biogas.

Gaseous oxygen is excluded from the reactions by physical containment. Anaerobes utilize electron acceptors from sources other than oxygen gas. These acceptors can be the organic material itself or may be supplied by inorganic oxides from within the input material. When the oxygen source in an anaerobic system is derived from the organic material itself, the 'intermediate' end products are primarily alcohols, aldehydes, and organic acids, plus carbon dioxide. In the presence of specialised methanogens, the intermediates are converted to the 'final' end products of methane, carbon dioxide, and trace levels of hydrogen sulfide. In an anaerobic system, the majority of the chemical energy contained within the starting material is released by methanogenic bacteria as methane.

Populations of anaerobic microorganisms typically take a significant period of time to establish themselves to be fully effective. Therefore, common practice is to introduce anaerobic microorganisms from materials with existing populations, a process known as "seeding" the digesters, typically accomplished with the addition of sewage sludge or cattle slurry.

Process Stages

The four key stages of anaerobic digestion involve hydrolysis, acidogenesis, acetogenesis and methanogenesis. The overall process can be described by the chemical reaction, where organic material such as glucose is biochemically digested into carbon dioxide (CO_2) and methane (CH_4) by the anaerobic microorganisms.

$$C_6H_{12}O_6 \rightarrow 3CO_2 + 3CH_4$$

- Hydrolysis:- In most cases, biomass is made up of large organic polymers. For the bacteria in anaerobic digesters to access the energy potential of the material, these chains must first be broken down into their smaller constituent parts. These constituent parts, or monomers, such as sugars, are readily available to other bacteria. The process of breaking these chains and dissolving the smaller molecules into solution is called hydrolysis. Therefore, hydrolysis of these high-molecular-weight polymeric components is the necessary first step in anaerobic digestion. Through hydrolysis the complex organic molecules are broken down into simple sugars, amino acids, and fatty acids.

  Acetate and hydrogen produced in the first stages can be used directly by methanogens. Other molecules, such as volatile fatty acids (VFAs) with a chain length greater than that of acetate must first be catabolised into compounds that can be directly used by methanogens.

- Acidogenesis:- The biological process of acidogenesis results in further breakdown of the remaining components by acidogenic (fermentative) bacteria. Here, VFAs are created, along with ammonia, carbon dioxide, and hydrogen sulfide, as well as other byproducts. The process of acidogenesis is similar to the way milk sours.

- Acetogenesis:- The third stage of anaerobic digestion is acetogenesis. Here, simple molecules created through the acidogenesis phase are further digested by acetogens to produce largely acetic acid, as well as carbon dioxide and hydrogen.

- Methanogenesis:- The terminal stage of anaerobic digestion is the biological process of methanogenesis. Here, methanogens use the intermediate products of the preceding stages and convert them into methane, carbon dioxide, and water. These components make up the majority of the biogas emitted from the system. Methanogenesis is sensitive to both high and low pHs and occurs between pH 6.5 and pH 8. The remaining, indigestible material the microbes cannot use and any dead bacterial remains constitute the digestate.

Configuration

Anaerobic digesters can be designed and engineered to operate using a number of different configurations and can be categorized into batch vs. continuous process mode, mesophilic vs. thermophilic temperature conditions, high vs. low portion of solids, and single stage vs. multistage processes. More initial build money and a larger volume of the batch digester is needed to handle the same amount of waste as a continuous process digester. Higher heat energy is demanded in a thermophilic system compared to a mesophilic system and has a larger gas output capacity and higher methane gas content. For solids content, low will handle up to 15% solid content. Above this level is considered high solids content and can also be known as dry digestion. In a single stage process, one reactor houses the four anaerobic digestion steps. A multistage process utilizes two or more reactors for digestion to separate the methanogenesis and hydrolysis phases.

Batch or Continuous

Anaerobic digestion can be performed as a batch process or a continuous process. In a batch system, biomass is added to the reactor at the start of the process. The reactor is then sealed for the duration of the process. In its simplest form batch processing needs inoculation with already processed material to start the anaerobic digestion. In a typical scenario, biogas production will be formed with a normal distribution pattern over time. Operators can use this fact to determine when they believe the process of digestion of the organic matter has completed. There can be severe odour issues if a batch reactor is opened and emptied before the process is well completed. A more advanced type of batch approach has limited the odour issues by integrating anaerobic digestion with in-vessel composting. In this approach inoculation takes place through the use of recirculated degasified percolate. After anaerobic digestion has completed, the biomass is kept in the reactor which is then used for in-vessel composting before it is opened As the batch digestion is simple and requires less equipment and lower levels of design work, it is typically a cheaper form of digestion. Using more than one batch reactor at a plant can ensure constant production of biogas.

In continuous digestion processes, organic matter is constantly added (continuous complete mixed) or added in stages to the reactor (continuous plug flow; first in – first out). Here, the end products are constantly or periodically removed, resulting in constant production of biogas. A single or multiple digesters in sequence may be used. Examples of this form of anaerobic digestion include continuous stirred-tank reactors, upflow anaerobic sludge blankets, expanded granular sludge beds and internal circulation reactors.

Temperature

The two conventional operational temperature levels for anaerobic digesters determine the species of methanogens in the digesters:

- *Mesophilic* digestion takes place optimally around 30 to 38 °C, or at ambient temperatures between 20 and 45 °C, where mesophiles are the primary microorganism present.
- *Thermophilic* digestion takes place optimally around 49 to 57 °C, or at elevated temperatures up to 70 °C, where thermophiles are the primary microorganisms present.

A limit case has been reached in Bolivia, with anaerobic digestion in temperature working conditions of less than 10 °C. The anaerobic process is very slow, taking more than three times the normal mesophilic time process. In experimental work at University of Alaska Fairbanks, a 1,000 litre digester using psychrophiles harvested from "mud from a frozen lake in Alaska" has produced 200–300 litres of methane per day, about 20 to 30% of the output from digesters in warmer climates. Mesophilic species outnumber thermophiles, and they are also more tolerant to changes in environmental conditions than thermophiles. Mesophilic systems are, therefore, considered to be more stable than thermophilic digestion systems. In contrast, while thermophilic digestion systems are considered less stable, their energy input is higher, with more biogas being removed from the organic matter in an equal amount of time. The increased temperatures facilitate faster reaction rates, and thus faster gas yields. Operation at higher temperatures facilitates greater pathogen reduction of the digestate. In countries where legislation, such as the Animal By-Products Regulations in the European Union, requires digestate to meet certain levels of pathogen reduction there may be a benefit to using thermophilic temperatures instead of mesophilic.

Additional pre-treatment can be used to reduce the necessary retention time to produce biogas. For example, certain processes shred the substrates to increase the surface area or use a thermal pretreatment stage (such as pasteurisation) to significantly enhance the biogas output. The pasteurisation process can also be used to reduce the pathogenic concentration in the digesate leaving the anaerobic digester. Pasteurisation may be achieved by heat treatment combined with maceration of the solids.

Solids Content

In a typical scenario, three different operational parameters are associated with the solids content of the feedstock to the digesters:

- High solids (dry—stackable substrate)
- High solids (wet—pumpable substrate)
- Low solids (wet—pumpable substrate)

High solids (dry) digesters are designed to process materials with a solids content between 25 and 40%. Unlike wet digesters that process pumpable slurries, high solids (dry – stackable substrate) digesters are designed to process solid substrates without the addition of water. The primary styles of dry digesters are continuous vertical plug flow and batch tunnel horizontal digesters. Continuous vertical plug flow digesters are upright, cylindrical tanks where feedstock is continuously fed into the top of the digester, and flows downward by gravity during digestion. In batch tunnel digesters, the feedstock is deposited in tunnel-like chambers with a gas-tight door. Neither approach has mixing inside the digester. The amount of pretreatment, such as contaminant removal, depends both upon the nature of the waste streams being processed and the desired quality of the digestate. Size reduction (grinding) is beneficial in continuous vertical systems, as it accelerates digestion, while batch systems avoid grinding and instead require structure (e.g. yard waste) to reduce compaction of the stacked pile. Continuous vertical dry digesters have a smaller footprint due to the shorter effective retention time and vertical design. Wet digesters can be designed to operate in either a high-solids content, with a total suspended solids (TSS) concentration greater than ~20%, or a low-solids concentration less than ~15%.

High solids (wet) digesters process a thick slurry that requires more energy input to move and process the feedstock. The thickness of the material may also lead to associated problems with abrasion. High solids digesters will typically have a lower land requirement due to the lower volumes associated with the moisture. High solids digesters also require correction of conventional performance calculations (e.g. gas production, retention time, kinetics, etc.) originally based on very dilute sewage digestion concepts, since larger fractions of the feedstock mass are potentially convertible to biogas.

Low solids (wet) digesters can transport material through the system using standard pumps that require significantly lower energy input. Low solids digesters require a larger amount of land than high solids due to the increased volumes associated with the increased liquid-to-feedstock ratio of the digesters. There are benefits associated with operation in a liquid environment, as it enables more thorough circulation of materials and contact between the bacteria and their food. This enables the bacteria to more readily access the substances on which they are feeding, and increases the rate of gas production.

Complexity

Digestion systems can be configured with different levels of complexity. In a single-stage digestion system (one-stage), all of the biological reactions occur within a single, sealed reactor or holding tank. Using a single stage reduces construction costs, but results in less control of the reactions occurring within the system. Acidogenic bacteria, through the production of acids, reduce the pH of the tank. Methanogenic bacteria, as outlined earlier, operate in a strictly defined pH range. Therefore, the biological reactions of the different species in a single-stage reactor can be in direct competition with each other. Another one-stage reaction system is an anaerobic lagoon. These lagoons are pond-like, earthen basins used for the treatment and long-term storage of manures. Here the anaerobic reactions are contained within the natural anaerobic sludge contained in the pool.

In a two-stage digestion system (multistage), different digestion vessels are optimised to bring maximum control over the bacterial communities living within the digesters. Acidogenic bacteria produce organic acids and more quickly grow and reproduce than methanogenic bacteria. Methanogenic bacteria require stable pH and temperature to optimise their performance.

Under typical circumstances, hydrolysis, acetogenesis, and acidogenesis occur within the first reaction vessel. The organic material is then heated to the required operational temperature (either mesophilic or thermophilic) prior to being pumped into a methanogenic reactor. The initial hydrolysis or acidogenesis tanks prior to the methanogenic reactor can provide a buffer to the rate at which feedstock is added. Some European countries require a degree of elevated heat treatment to kill harmful bacteria in the input waste. In this instance, there may be a pasteurisation or sterilisation stage prior to digestion or between the two digestion tanks. Notably, it is not possible to completely isolate the different reaction phases, and often some biogas is produced in the hydrolysis or acidogenesis tanks.

Residence Time

The residence time in a digester varies with the amount and type of feed material, and with the configuration of the digestion system. In a typical two-stage mesophilic digestion, residence time varies between 15 and 40 days, while for a single-stage thermophilic digestion, residence times is normally faster and takes around 14 days. The plug-flow nature of some of these systems will mean the full degradation of the material may not have been realised in this timescale. In this event, digestate exiting the system will be darker in colour and will typically have more odour.

In the case of an upflow anaerobic sludge blanket digestion (UASB), hydraulic residence times can be as short as 1 hour to 1 day, and solid retention times can be up to 90 days. In this manner, a UASB system is able to separate solids and hydraulic retention times with the use of a sludge blanket. Continuous digesters have mechanical or hydraulic devices, depending on the level of solids in the material, to mix the contents, enabling the bacteria and the food to be in contact. They also allow excess material to be continuously extracted to maintain a reasonably constant volume within the digestion tanks.

Inhibition

The anaerobic digestion process can be inhibited by several compounds, affecting one or more of the bacterial groups responsible for the different organic matter degradation steps. The degree of the inhibition depends, among other factors, on the concentration of the inhibitor in the digester. Potential inhibitors are ammonia, sulfide, light metal ions (Na, K, Mg, Ca, Al), heavy metals, some organics (chlorophenols, halogenated aliphatics, N-substituted aromatics, long chain fatty acids), etc.

Left: Farm-based maize silage digester located near Neumünster in Germany, 2007 - the green, inflatable biogas holder is shown on top of the digester. *Right:* Two-stage, low solids, UASB digestion component of a mechanical biological treatment system near Tel Aviv; the process water is seen in balance tank and sequencing batch reactor, 2005.

Feedstocks

Anaerobic lagoon and generators at the Cal Poly Dairy, United States

The most important initial issue when considering the application of anaerobic digestion systems is the feedstock to the process. Almost any organic material can be processed with anaerobic digestion; however, if biogas production is the aim, the level of putrescibility is the key factor in its successful application. The more putrescible (digestible) the material, the higher the gas yields possible from the system.

Feedstocks can include biodegradable waste materials, such as waste paper, grass clippings, leftover food, sewage, and animal waste. Woody wastes are the exception, because they are largely unaffected by digestion, as most anaerobes are unable to degrade lignin. Xylophalgeous anaerobes (lignin consumers) or using high temperature pretreatment, such as pyrolysis, can be used to break down the lignin. Anaerobic digesters can also be fed with specially grown energy crops, such as silage, for dedicated biogas production. In Germany and continental Europe, these facilities are referred to as "biogas" plants. A codigestion or cofermentation plant is typically an agricultural anaerobic digester that accepts two or more input materials for simultaneous digestion.

The length of time required for anaerobic digestion depends on the chemical complexity of the material. Material rich in easily digestible sugars breaks down quickly where as intact lignocellulosic material rich in cellulose and hemicellulose polymers can take much longer to break down. Anaerobic microorganisms are generally unable to break down lignin, the recalcitrant aromatic component of biomass.

Anaerobic digesters were originally designed for operation using sewage sludge and manures. Sewage and manure are not, however, the material with the most potential for anaerobic digestion, as the biodegradable material has already had much of the energy content taken out by the animals that produced it. Therefore, many digesters operate with codigestion of two or more types of feedstock. For example, in a farm-based digester that uses dairy manure as the primary feedstock, the gas production may be significantly increased by adding a second feedstock, e.g., grass and corn (typical on-farm feedstock), or various organic byproducts, such as slaughterhouse waste, fats, oils and grease from restaurants, organic household waste, etc. (typical off-site feedstock).

Digesters processing dedicated energy crops can achieve high levels of degradation and biogas production. Slurry-only systems are generally cheaper, but generate far less energy than those using crops, such as maize and grass silage; by using a modest amount of crop material (30%), an anaerobic digestion plant can increase energy output tenfold for only three times the capital cost, relative to a slurry-only system.

Moisture Content

A second consideration related to the feedstock is moisture content. Dryer, stackable substrates, such as food and yard waste, are suitable for digestion in tunnel-like chambers. Tunnel-style systems typically have near-zero wastewater discharge, as well, so this style of system has advantages where the discharge of digester liquids are a liability. The wetter the material, the more suitable it will be to handling with standard pumps instead of energy-intensive concrete pumps and physical means of movement. Also, the wetter the material, the more volume and area it takes up relative to the levels of gas produced. The moisture content of the target feedstock will also affect what type of system is applied to its treatment. To use a high-solids anaerobic digester for dilute feedstocks, bulking agents, such as compost, should be applied to increase the solids content of the input material. Another key consideration is the carbon:nitrogen ratio of the input material. This ratio is the balance of food a microbe requires to grow; the optimal C:N ratio is 20–30:1. Excess N can lead to ammonia inhibition of digestion.

Contamination

The level of contamination of the feedstock material is a key consideration. If the feedstock to the digesters has significant levels of physical contaminants, such as plastic, glass, or metals, then processing to remove the contaminants will be required for the material to be used. If it is not removed, then the digesters can be blocked and will not function efficiently. It is with this understanding that mechanical biological treatment plants are designed. The higher the level of pretreatment a feedstock requires, the more processing machinery will be required, and, hence, the project will have higher capital costs.

After sorting or screening to remove any physical contaminants from the feedstock, the material is often shredded, minced, and mechanically or hydraulically pulped to increase the surface area available to microbes in the digesters and, hence, increase the speed of digestion. The maceration of solids can be achieved by using a chopper pump to transfer the feedstock material into the airtight digester, where anaerobic treatment takes place.

Substrate Composition

Substrate composition is a major factor in determining the methane yield and methane production rates from the digestion of biomass. Techniques to determine the compositional characteristics of the feedstock are available, while parameters such as solids, elemental, and organic analyses are important for digester design and operation.

Applications

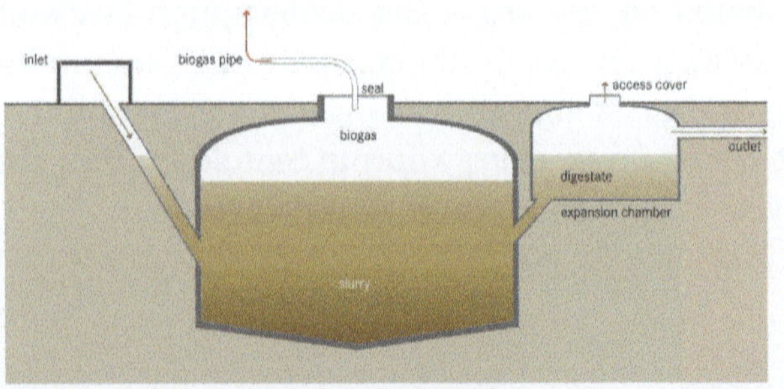

Schematic of an anaerobic digester as part of a sanitation system. It produces a digested slurry (digestate) that can be used as a fertilizer, and biogas that can be used for energy.

Using anaerobic digestion technologies can help to reduce the emission of greenhouse gases in a number of key ways:

- Replacement of fossil fuels
- Reducing or eliminating the energy footprint of waste treatment plants
- Reducing methane emission from landfills
- Displacing industrially produced chemical fertilizers
- Reducing vehicle movements

- Reducing electrical grid transportation losses
- Reducing usage of LP Gas for cooking

Waste and Wastewater Treatment

Anaerobic digestion is particularly suited to organic material, and is commonly used for industrial effluent, wastewater and sewage sludge treatment. Anaerobic digestion, a simple process, can greatly reduce the amount of organic matter which might otherwise be destined to be dumped at sea, dumped in landfills, or burnt in incinerators.

Pressure from environmentally related legislation on solid waste disposal methods in developed countries has increased the application of anaerobic digestion as a process for reducing waste volumes and generating useful byproducts. It may either be used to process the source-separated fraction of municipal waste or alternatively combined with mechanical sorting systems, to process residual mixed municipal waste. These facilities are called mechanical biological treatment plants.

If the putrescible waste processed in anaerobic digesters were disposed of in a landfill, it would break down naturally and often anaerobically. In this case, the gas will eventually escape into the atmosphere. As methane is about 20 times more potent as a greenhouse gas than carbon dioxide, this has significant negative environmental effects.

In countries that collect household waste, the use of local anaerobic digestion facilities can help to reduce the amount of waste that requires transportation to centralized landfill sites or incineration facilities. This reduced burden on transportation reduces carbon emissions from the collection vehicles. If localized anaerobic digestion facilities are embedded within an electrical distribution network, they can help reduce the electrical losses associated with transporting electricity over a national grid.

Power Generation

In developing countries, simple home and farm-based anaerobic digestion systems offer the potential for low-cost energy for cooking and lighting. From 1975, China and India have both had large, government-backed schemes for adaptation of small biogas plants for use in the household for cooking and lighting. At present, projects for anaerobic digestion in the developing world can gain financial support through the United Nations Clean Development Mechanism if they are able to show they provide reduced carbon emissions.

Methane and power produced in anaerobic digestion facilities can be used to replace energy derived from fossil fuels, and hence reduce emissions of greenhouse gases, because the carbon in biodegradable material is part of a carbon cycle. The carbon released into the atmosphere from the combustion of biogas has been removed by plants for them to grow in the recent past, usually within the last decade, but more typically within the last growing season. If the plants are regrown, taking the carbon out of the atmosphere once more, the system will be carbon neutral. In contrast, carbon in fossil fuels has been sequestered in the earth for many millions of years, the combustion of which increases the overall levels of carbon dioxide in the atmosphere.

Biogas from sewage sludge treatment is sometimes used to run a gas engine to produce electrical power, some or all of which can be used to run the sewage works. Some waste heat from the engine is then used to heat the digester. The waste heat is, in general, enough to heat the digester to the required temperatures. The power potential from sewage works is limited – in the UK, there are about 80 MW total of such generation, with the potential to increase to 150 MW, which is insignificant compared to the average power demand in the UK of about 35,000 MW. The scope for biogas generation from nonsewage waste biological matter – energy crops, food waste, abattoir waste, etc. - is much higher, estimated to be capable of about 3,000 MW. Farm biogas plants using animal waste and energy crops are expected to contribute to reducing CO_2 emissions and strengthen the grid, while providing UK farmers with additional revenues.

Some countries offer incentives in the form of, for example, feed-in tariffs for feeding electricity onto the power grid to subsidize green energy production.

In Oakland, California at the East Bay Municipal Utility District's main wastewater treatment plant (EBMUD), food waste is currently codigested with primary and secondary municipal wastewater solids and other high-strength wastes. Compared to municipal wastewater solids digestion alone, food waste codigestion has many benefits. Anaerobic digestion of food waste pulp from the EBMUD food waste process provides a higher normalized energy benefit, compared to municipal wastewater solids: 730 to 1,300 kWh per dry ton of food waste applied compared to 560 to 940 kWh per dry ton of municipal wastewater solids applied.

Grid Injection

Biogas grid-injection is the injection of biogas into the natural gas grid. The raw biogas has to be previously upgraded to biomethane. This upgrading implies the removal of contaminants such as hydrogen sulphide or siloxanes, as well as the carbon dioxide. Several technologies are available for this purpose, being the most widely implemented the pressure swing adsorption (PSA), water or amine scrubbing (absorption processes) and, in the last years, membrane separation. As an alternative, the electricity and the heat can be used for on-site generation, resulting in a reduction of losses in the transportation of energy. Typical energy losses in natural gas transmission systems range from 1–2%, whereas the current energy losses on a large electrical system range from 5–8%.

In October 2010, Didcot Sewage Works became the first in the UK to produce biomethane gas supplied to the national grid, for use in up to 200 homes in Oxfordshire. By 2017, UK electricity firm Ecotricity plan to have digester fed by locally sourced grass fueling 6000 homes

Vehicle Fuel

After upgrading with the above-mentioned technologies, the biogas (transformed into biomethane) can be used as vehicle fuel in adapted vehicles. This use is very extensive in Sweden, where over 38,600 gas vehicles exist, and 60% of the vehicle gas is biomethane generated in anaerobic digestion plants.

Fertiliser and Soil Conditioner

The solid, fibrous component of the digested material can be used as a soil conditioner to increase the organic content of soils. Digester liquor can be used as a fertiliser to supply vital nutrients to soils instead of chemical fertilisers that require large amounts of energy to produce and transport. The use of manufactured fertilisers is, therefore, more carbon-intensive than the use of anaerobic digester liquor fertiliser. In countries such as Spain, where many soils are organically depleted, the markets for the digested solids can be equally as important as the biogas.

Cooking Gas

By using a bio-digester, which produces the bacteria required for decomposing, cooking gas is generated. The organic garbage like fallen leaves, kitchen waste, food waste etc. are fed into a crusher unit, where the mixture is conflated with a small amount of water. The mixture is then fed into the bio-digester, where the bacteria decomposes it to produce cooking gas. This gas is piped to kitchen stove. A 2 cubic meter bio-digester can produce 2 cubic meter of cooking gas. This is equivalent to 1 kg of LPG. The notable advantage of using a bio-digester is the sludge which is a rich organic manure.

Products

The three principal products of anaerobic digestion are biogas, digestate, and water.

Biogas

Typical composition of biogas		
Compound	**Formula**	**%**
Methane	CH_4	50–75
Carbon dioxide	CO_2	25–50
Nitrogen	N_2	0–10
Hydrogen	H_2	0–1
Hydrogen sulfide	H_2S	0–3
Oxygen	O_2	0–0

Biogas is the ultimate waste product of the bacteria feeding off the input biodegradable feedstock (the methanogenesis stage of anaerobic digestion is performed by archaea (a micro-organism on a distinctly different branch of the phylogenetic tree of life to bacteria), and is mostly methane and carbon dioxide, with a small amount hydrogen and trace hydrogen sulfide. (As-produced, biogas also contains water vapor, with the fractional water vapor volume a function of biogas temperature). Most of the biogas is produced during the middle of the digestion, after the bacterial population has grown, and tapers off as the putrescible material is exhausted. The gas is normally stored on top of the digester in an inflatable gas bubble or extracted and stored next to the facility in a gas holder.

The methane in biogas can be burned to produce both heat and electricity, usually with a reciprocating engine or microturbine often in a cogeneration arrangement where the electricity and waste heat generated are used to warm the digesters or to heat buildings. Excess electricity can be sold to suppliers or put into the local grid. Electricity produced by anaerobic digesters is considered to be renewable energy and may attract subsidies. Biogas does not contribute to increasing atmospheric carbon dioxide concentrations because the gas is not released directly into the atmosphere and the carbon dioxide comes from an organic source with a short carbon cycle.

Biogas may require treatment or 'scrubbing' to refine it for use as a fuel. Hydrogen sulfide, a toxic product formed from sulfates in the feedstock, is released as a trace component of the biogas. National environmental enforcement agencies, such as the U.S. Environmental Protection Agency or the English and Welsh Environment Agency, put strict limits on the levels of gases containing hydrogen sulfide, and, if the levels of hydrogen sulfide in the gas are high, gas scrubbing and cleaning equipment (such as amine gas treating) will be needed to process the biogas to within regionally accepted levels. Alternatively, the addition of ferrous chloride $FeCl_2$ to the digestion tanks inhibits hydrogen sulfide production.

Volatile siloxanes can also contaminate the biogas; such compounds are frequently found in household waste and wastewater. In digestion facilities accepting these materials as a component of the feedstock, low-molecular-weight siloxanes volatilise into biogas. When this gas is combusted in a gas engine, turbine, or boiler, siloxanes are converted into silicon dioxide (SiO_2), which deposits internally in the machine, increasing wear and tear. Practical and cost-effective technologies to remove siloxanes and other biogas contaminants are available at the present time. In certain applications, *in situ* treatment can be used to increase the methane purity by reducing the offgas carbon dioxide content, purging the majority of it in a secondary reactor.

In countries such as Switzerland, Germany, and Sweden, the methane in the biogas may be compressed for it to be used as a vehicle transportation fuel or input directly into the gas mains. In countries where the driver for the use of anaerobic digestion are renewable electricity subsidies, this route of treatment is less likely, as energy is required in this processing stage and reduces the overall levels available to sell.

Biogas holder with lightning protection rods and backup gas flare

Biogas carrying pipes

Digestate

Digestate is the solid remnants of the original input material to the digesters that the microbes cannot use. It also consists of the mineralised remains of the dead bacteria from within the digesters. Digestate can come in three forms: fibrous, liquor, or a sludge-based combination of the two fractions. In two-stage systems, different forms of digestate come from different digestion tanks. In single-stage digestion systems, the two fractions will be combined and, if desired, separated by further processing.

Acidogenic anaerobic digestate

The second byproduct (acidogenic digestate) is a stable, organic material consisting largely of lignin and cellulose, but also of a variety of mineral components in a matrix of dead bacterial cells; some plastic may be present. The material resembles domestic compost and can be used as such or to make low-grade building products, such as fibreboard. The solid digestate can also be used as feedstock for ethanol production.

The third byproduct is a liquid (methanogenic digestate) rich in nutrients, which can be used as a fertiliser, depending on the quality of the material being digested. Levels of potentially toxic elements (PTEs) should be chemically assessed. This will depend upon the quality of the original feedstock. In the case of most clean and source-separated biodegradable waste streams, the levels of PTEs will be low. In the case of wastes originating from industry, the levels of PTEs may be higher and will need to be taken into consideration when determining a suitable end use for the material.

Digestate typically contains elements, such as lignin, that cannot be broken down by the anaerobic microorganisms. Also, the digestate may contain ammonia that is phytotoxic, and may hamper the growth of plants if it is used as a soil-improving material. For these two reasons, a maturation or composting stage may be employed after digestion. Lignin and other materials are available for degradation by aerobic microorganisms, such as fungi, helping reduce the overall volume of the material for transport. During this maturation, the ammonia will be oxidized into nitrates, improving the fertility of the material and making it more suitable as a soil improver. Large composting stages are typically used by dry anaerobic digestion technologies.

Wastewater

The final output from anaerobic digestion systems is water, which originates both from the moisture content of the original waste that was treated and water produced during the microbial reactions in the digestion systems. This water may be released from the dewatering of the digestate or may be implicitly separate from the digestate.

The wastewater exiting the anaerobic digestion facility will typically have elevated levels of biochemical oxygen demand (BOD) and chemical oxygen demand (COD). These measures of the reactivity of the effluent indicate an ability to pollute. Some of this material is termed 'hard COD', meaning it cannot be accessed by the anaerobic bacteria for conversion into biogas. If this effluent were put directly into watercourses, it would negatively affect them by causing eutrophication. As such, further treatment of the wastewater is often required. This treatment will typically be an oxidation stage wherein air is passed through the water in a sequencing batch reactors or reverse osmosis unit.

History

Gas street lamp

Reported scientific interest in the manufacturing of gas produced by the natural decomposition of organic matter dates from the 17th century, when Robert Boyle (1627-1691) and Stephen Hales (1677-1761) noted that disturbing the sediment of streams and lakes released flammable gas. In 1808 Sir Humphry Davy proved the presence of methane in the gases produced by cattle manure. In 1859 a leper colony in Bombay in India built the first anaerobic digester. In 1895, the technology was developed in Exeter, England, where a septic tank was used to generate gas for the sewer gas destructor lamp, a type of gas lighting. Also in England, in 1904, the first dual-purpose tank for both sedimentation and sludge treatment was installed in Hampton, London. In 1907, in Germany, a patent was issued for the Imhoff tank, an early form of digester.

Research on anaerobic digestion began in earnest in the 1930s.

Biofuel

A bus fueled by biodiesel

Information on pump regarding ethanol fuel blend up to 10%, California

A biofuel is a fuel that is produced through contemporary biological processes, such as agriculture and anaerobic digestion, rather than a fuel produced by geological processes such as those involved in the formation of fossil fuels, such as coal and petroleum, from prehistoric biological matter.

Biofuels can be derived directly from plants, or indirectly from agricultural, commercial, domestic, and/or industrial wastes. Renewable biofuels generally involve contemporary carbon fixation, such as those that occur in plants or microalgae through the process of photosynthesis. Other renewable biofuels are made through the use or conversion of biomass (referring to recently living organisms, most often referring to plants or plant-derived materials). This biomass can be converted to convenient energy-containing substances in three different ways: thermal conversion, chemical conversion, and biochemical conversion. This biomass conversion can result in fuel in solid, liquid, or gas form. This new biomass can also be used directly for biofuels.

Bioethanol is an alcohol made by fermentation, mostly from carbohydrates produced in sugar or starch crops such as corn, sugarcane, or sweet sorghum. Cellulosic biomass, derived from non-food sources, such as trees and grasses, is also being developed as a feedstock for ethanol production. Ethanol can be used as a fuel for vehicles in its pure form, but it is usually used as a gasoline additive to increase octane and improve vehicle emissions. Bioethanol is widely used in the USA and in Brazil. Current plant design does not provide for converting the lignin portion of plant raw materials to fuel components by fermentation.

Biodiesel can be used as a fuel for vehicles in its pure form, but it is usually used as a diesel additive to reduce levels of particulates, carbon monoxide, and hydrocarbons from diesel-powered vehicles. Biodiesel is produced from oils or fats using transesterification and is the most common biofuel in Europe.

In 2010, worldwide biofuel production reached 105 billion liters (28 billion gallons US), up 17% from 2009, and biofuels provided 2.7% of the world's fuels for road transport. Global ethanol fuel production reached 86 billion liters (23 billion gallons US) in 2010, with the United States and Brazil as the world's top producers, accounting together for about 90% of global production. The world's largest biodiesel producer is the European Union, accounting for 53% of all biodiesel production in 2010. As of 2011, mandates for blending biofuels exist in 31 countries at the national level and in 29 states or provinces. The International Energy Agency has a goal for biofuels to meet more than a quarter of world demand for transportation fuels by 2050 to reduce dependence on petroleum and coal. The production of biofuels also led into a flourishing automotive industry, where by 2010, 79% of all cars produced in Brazil were made with a hybrid fuel system of bioethanol and gasoline.

There are various social, economic, environmental and technical issues relating to biofuels production and use, which have been debated in the popular media and scientific journals. These include: the effect of moderating oil prices, the "food vs fuel" debate, poverty reduction potential, carbon emissions levels, sustainable biofuel production, deforestation and soil erosion, loss of biodiversity, impact on water resources, rural social exclusion and injustice, shantytown migration, rural unskilled unemployment, and nitrogen dioxide (NO_2) emissions.

Liquid Fuels for Transportation

Most transportation fuels are liquids, because vehicles usually require high energy density. This occurs naturally in liquids and solids. High energy density can also be provided by an internal combustion engine. These engines require clean-burning fuels. The fuels that are easiest to burn cleanly are typically liquids and gases. Thus, liquids meet the requirements of being both energy-dense and clean-burning. In addition, liquids (and gases) can be pumped, which means handling is easily mechanized, and thus less laborious.

First-generation Biofuels

"First-generation" or conventional biofuels are made from sugar, starch, or vegetable oil.

Ethanol

Neat ethanol on the left (A), gasoline on the right (G) at a filling station in Brazil

Biologically produced alcohols, most commonly ethanol, and less commonly propanol and butanol, are produced by the action of microorganisms and enzymes through the fermentation of sugars or starches (easiest), or cellulose (which is more difficult). Biobutanol (also called biogasoline) is often claimed to provide a direct replacement for gasoline, because it can be used directly in a gasoline engine.

U.S. President George W. Bush looks at sugar cane, a source of biofuel, with Brazilian President Luiz Inácio Lula da Silva during a tour on biofuel technology at Petrobras in São Paulo, Brazil, 9 March 2007.

Ethanol fuel is the most common biofuel worldwide, particularly in Brazil. Alcohol fuels are produced by fermentation of sugars derived from wheat, corn, sugar beets, sugar cane, molasses and any sugar or starch from which alcoholic beverages such as whiskey, can be made (such as potato and fruit waste, etc.). The ethanol production methods used are enzyme digestion (to release sugars from stored starches), fermentation of the sugars, distillation and drying. The distillation process requires significant energy input for heat (sometimes unsustainable natural gas fossil fuel, but cellulosic biomass such as bagasse, the waste left after sugar cane is pressed to extract its juice, is the most common fuel in Brazil, while pellets, wood chips and also waste heat are more common in Europe) Waste steam fuels ethanol factory – where waste heat from the factories also is used in the district heating grid.

Ethanol can be used in petrol engines as a replacement for gasoline; it can be mixed with gasoline to any percentage. Most existing car petrol engines can run on blends of up to 15% bioethanol with petroleum/gasoline. Ethanol has a smaller energy density than that of gasoline; this means it takes more fuel (volume and mass) to produce the same amount of work. An advantage of ethanol (CH_3CH_2OH) is that it has a higher octane rating than ethanol-free gasoline available at roadside gas stations, which allows an increase of an engine's compression ratio for increased thermal efficiency. In high-altitude (thin air) locations, some states mandate a mix of gasoline and ethanol as a winter oxidizer to reduce atmospheric pollution emissions.

Ethanol is also used to fuel bioethanol fireplaces. As they do not require a chimney and are "flue-less", bioethanol fires are extremely useful for newly built homes and apartments without a flue. The downsides to these fireplaces is that their heat output is slightly less than electric heat or gas fires, and precautions must be taken to avoid carbon monoxide poisoning.

Corn-to-ethanol and other food stocks has led to the development of cellulosic ethanol. According to a joint research agenda conducted through the US Department of Energy, the fossil energy ratios (FER) for cellulosic ethanol, corn ethanol, and gasoline are 10.3, 1.36, and 0.81, respectively.

Ethanol has roughly one-third lower energy content per unit of volume compared to gasoline. This is partly counteracted by the better efficiency when using ethanol (in a long-term test of more than 2.1 million km, the BEST project found FFV vehicles to be 1-26 % more energy efficient than petrol cars, but the volumetric consumption increases by approximately 30%, so more fuel stops are required).

With current subsidies, ethanol fuel is slightly cheaper per distance traveled in the United States.

Biodiesel

Biodiesel is the most common biofuel in Europe. It is produced from oils or fats using transesterification and is a liquid similar in composition to fossil/mineral diesel. Chemically, it consists mostly of fatty acid methyl (or ethyl) esters (FAMEs). Feedstocks for biodiesel include animal fats, vegetable oils, soy, rapeseed, jatropha, mahua, mustard, flax, sunflower, palm oil, hemp, field pennycress, *Pongamia pinnata* and algae. Pure biodiesel (B100, also known as "neat" biodiesel) currently reduces emissions with up to 60% compared to diesel Second generation B100.

Biodiesel can be used in any diesel engine when mixed with mineral diesel. In some countries, manufacturers cover their diesel engines under warranty for B100 use, although Volkswagen of Germany, for example, asks drivers to check by telephone with the VW environmental services department before switching to B100. B100 may become more viscous at lower temperatures, depending on the feedstock used. In most cases, biodiesel is compatible with diesel engines from 1994 onwards, which use 'Viton' (by DuPont) synthetic rubber in their mechanical fuel injection systems. Note however, that no vehicles are certified for using pure biodiesel before 2014, as there was no emission control protocol available for biodiesel before this date.

Electronically controlled 'common rail' and 'unit injector' type systems from the late 1990s onwards may only use biodiesel blended with conventional diesel fuel. These engines have finely metered and atomized multiple-stage injection systems that are very sensitive to the viscosity of the fuel. Many current-generation diesel engines are made so that they can run on B100 without altering the engine itself, although this depends on the fuel rail design. Since biodiesel is an effective solvent and cleans residues deposited by mineral diesel, engine filters may need to be replaced more often, as the biofuel dissolves old deposits in the fuel tank and pipes. It also effectively cleans the engine combustion chamber of carbon deposits, helping to maintain efficiency. In many European countries, a 5% biodiesel blend is widely used and is available at thousands of gas stations. Biodiesel is also an oxygenated fuel, meaning it contains a reduced amount of carbon and higher hydrogen and oxygen content than fossil diesel. This improves the combustion of biodiesel and reduces the particulate emissions from unburnt carbon. However, using pure biodiesel may increase NO_x-emissions.

Biodiesel is also safe to handle and transport because it is non-toxic and biodegradable, and has a high flash point of about 300 °F (148 °C) compared to petroleum diesel fuel, which has a flash point of 125 °F (52 °C).

In the USA, more than 80% of commercial trucks and city buses run on diesel. The emerging US biodiesel market is estimated to have grown 200% from 2004 to 2005. "By the end of 2006 biodiesel production was estimated to increase fourfold [from 2004] to more than" 1 billion US gallons (3,800,000 m^3).

In France, biodiesel is incorporated at a rate of 8% in the fuel used by all French diesel vehicles. Avril Group produces under the brand Diester, a fifth of 11 million tons of biodiesel consumed annually by the European Union. It is the leading European producer of biodiesel.

Other Bioalcohols

Methanol is currently produced from natural gas, a non-renewable fossil fuel. In the future it is hoped to be produced from biomass as biomethanol. This is technically feasible, but the production is currently being postponed for concerns of Jacob S. Gibbs and Brinsley Coleberd that the economic viability is still pending. The methanol economy is an alternative to the hydrogen economy, compared to today's hydrogen production from natural gas.

Butanol (C_4H_9OH) is formed by ABE fermentation (acetone, butanol, ethanol) and experimental modifications of the process show potentially high net energy gains with butanol as the only liquid product. Butanol will produce more energy and allegedly can be burned "straight" in existing gasoline engines (without modification to the engine or car), and is less corrosive and less water-soluble than ethanol, and could be distributed via existing infrastructures. DuPont and BP are working together to help develop butanol. *E. coli* strains have also been successfully engineered to produce butanol by modifying their amino acid metabolism.

Green Diesel

Green diesel is produced through hydrocracking biological oil feedstocks, such as vegetable oils and animal fats. Hydrocracking is a refinery method that uses elevated temperatures and pressure in the presence of a catalyst to break down larger molecules, such as those found in vegetable oils, into shorter hydrocarbon chains used in diesel engines. It may also be called renewable diesel, hydrotreated vegetable oil or hydrogen-derived renewable diesel. Green diesel has the same chemical properties as petroleum-based diesel. It does not require new engines, pipelines or infrastructure to distribute and use, but has not been produced at a cost that is competitive with petroleum. Gasoline versions are also being developed. Green diesel is being developed in Louisiana and Singapore by ConocoPhillips, Neste Oil, Valero, Dynamic Fuels, and Honeywell UOP as well as Preem in Gothenburg, Sweden, creating what is known as Evolution Diesel.

Biofuel Gasoline

In 2013 UK researchers developed a genetically modified strain of Escherichia coli (E.Coli), which could transform glucose into biofuel gasoline that does not need to be blended. Later in 2013 UCLA researchers engineered a new metabolic pathway to bypass glycolysis and increase the rate of conversion of sugars into biofuel, while KAIST researchers developed a strain capable of producing short-chain alkanes, free fatty acids, fatty esters and fatty alcohols through the fatty acyl (acyl carrier protein (ACP)) to fatty acid to fatty acyl-CoA pathway *in vivo*. It is believed that in the future it will be possible to "tweak" the genes to make gasoline from straw or animal manure.

Vegetable Oil

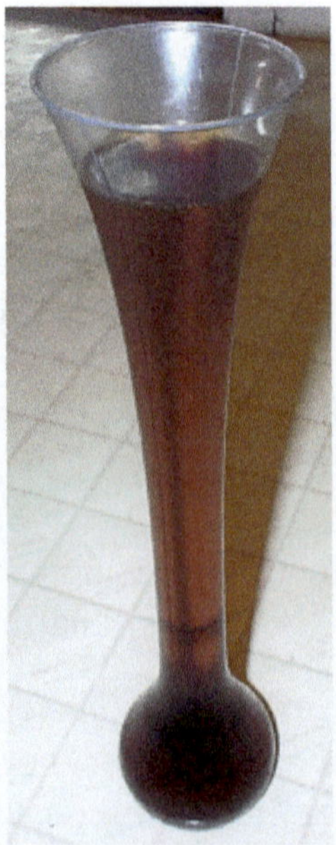

Filtered waste vegetable oil

Walmart's truck fleet logs millions of miles each year, and the company planned to double the fleet's efficiency between 2005 and 2015. This truck is one of 15 based at Walmart's Buckeye, Arizona distribution center that was converted to run on a biofuel made from reclaimed cooking grease produced during food preparation at Walmart stores.

Straight unmodified edible vegetable oil is generally not used as fuel, but lower-quality oil has been used for this purpose. Used vegetable oil is increasingly being processed into biodiesel, or (more rarely) cleaned of water and particulates and then used as a fuel.

As with 100% biodiesel (B100), to ensure the fuel injectors atomize the vegetable oil in the correct pattern for efficient combustion, vegetable oil fuel must be heated to reduce its viscosity to that of diesel, either by electric coils or heat exchangers. This is easier in warm or temperate climates. MAN B&W Diesel, Wärtsilä, and Deutz AG, as well as a number of smaller companies, such as Elsbett, offer engines that are compatible with straight vegetable oil, without the need for after-market modifications.

Vegetable oil can also be used in many older diesel engines that do not use common rail or unit injection electronic diesel injection systems. Due to the design of the combustion chambers in indirect injection engines, these are the best engines for use with vegetable oil. This system allows the relatively larger oil molecules more time to burn. Some older engines, especially Mercedes, are driven experimentally by enthusiasts without any conversion, a handful of drivers have experienced limited success with earlier pre-"Pumpe Duse" VW TDI engines and other similar engines with direct injection. Several companies, such as Elsbett or Wolf, have developed professional conversion kits and successfully installed hundreds of them over the last decades.

Oils and fats can be hydrogenated to give a diesel substitute. The resulting product is a straight-chain hydrocarbon with a high cetane number, low in aromatics and sulfur and does not contain oxygen. Hydrogenated oils can be blended with diesel in all proportions. They have several advantages over biodiesel, including good performance at low temperatures, no storage stability problems and no susceptibility to microbial attack.

Bioethers

Bioethers (also referred to as fuel ethers or oxygenated fuels) are cost-effective compounds that act as octane rating enhancers."Bioethers are produced by the reaction of reactive iso-olefins, such as iso-butylene, with bioethanol." Bioethers are created by wheat or sugar beet. They also enhance engine performance, whilst significantly reducing engine wear and toxic exhaust emissions. Though bioethers are likely to replace petroethers in the UK, it is highly unlikely they will become a fuel in and of itself due to the low energy density. Greatly reducing the amount of ground-level ozone emissions, they contribute to air quality.

When it comes to transportation fuel there are six ether additives: dimethyl ether (DME), diethyl ether (DEE), methyl teritiary-butyl ether (MTBE), ethyl *ter*-butyl ether (ETBE), *ter*-amyl methyl ether (TAME), and *ter*-amyl ethyl ether (TAEE).

The European Fuel Oxygenates Association (EFOA) credits methyl Ttertiary-butyl ether (MTBE) and ethyl ter-butyl ether (ETBE) as the most commonly used ethers in fuel to replace lead. Ethers were introduced in Europe in the 1970s to replace the highly toxic compound. Although Europeans still use bio-ether additives, the US no longer has an oxygenate requirement therefore bio-ethers are no longer used as the main fuel additive.

Biogas

Biogas is methane produced by the process of anaerobic digestion of organic material by anaerobes. It can be produced either from biodegradable waste materials or by the use of energy crops fed into anaerobic digesters to supplement gas yields. The solid byproduct, digestate, can be used as a biofuel or a fertilizer.

Biogas can be recovered from mechanical biological treatment waste processing systems. Landfill gas, a less clean form of biogas, is produced in landfills through naturally occurring anaerobic digestion. If it escapes into the atmosphere, it is a potential greenhouse gas.

Farmers can produce biogas from manure from their cattle by using anaerobic digesters.

Syngas

Syngas, a mixture of carbon monoxide, hydrogen and other hydrocarbons, is produced by partial combustion of biomass, that is, combustion with an amount of oxygen that is not sufficient to convert the biomass completely to carbon dioxide and water. Before partial combustion, the biomass is dried, and sometimes pyrolysed. The resulting gas mixture, syngas, is more efficient than direct combustion of the original biofuel; more of the energy contained in the fuel is extracted.

Syngas may be burned directly in internal combustion engines, turbines or high-temperature fuel cells. The wood gas generator, a wood-fueled gasification reactor, can be connected to an internal combustion engine.

Syngas can be used to produce methanol, DME and hydrogen, or converted via the Fischer-Tropsch process to produce a diesel substitute, or a mixture of alcohols that can be blended into gasoline. Gasification normally relies on temperatures greater than 700 °C.

Lower-temperature gasification is desirable when co-producing biochar, but results in syngas polluted with tar.

Solid Biomass Fuels

Examples include wood, sawdust, grass trimmings, domestic refuse, charcoal, agricultural waste, nonfood energy crops, and dried manure.

When solid biomass is already in a suitable form (such as firewood), it can burn directly in a stove or furnace to provide heat or raise steam. When solid biomass is in an inconvenient form (such as sawdust, wood chips, grass, urban waste wood, agricultural residues), the typical process is to densify the biomass. This process includes grinding the raw biomass to an appropriate particulate size (known as hogfuel), which, depending on the densification type, can be from 1 to 3 cm (0.4 to 1.2 in), which is then concentrated into a fuel product. The current processes produce wood pellets, cubes, or pucks. The pellet process is most common in Europe, and is typically a pure wood product. The other types of densification are larger in size compared to a pellet and are compatible with a broad range of input feedstocks. The resulting densified fuel is easier to transport and feed into thermal generation systems, such as boilers.

Sawdust, bark and chips are already used for decades for fuel in industrial processes; examples include the pulp and paper industry and the sugar cane industry. Boilers in the range of 500,000 lb/hr of steam, and larger, are in routine operation, using grate, spreader stoker, suspension burning and fluid bed combustion. Utilities generate power, typically in the range of 5 to 50 MW, using locally available fuel. Other industries have also installed wood waste fueled boilers and dryers in areas with low-cost fuel.

One of the advantages of solid biomass fuel is that it is often a byproduct, residue or waste-product of other processes, such as farming, animal husbandry and forestry. In theory, this means fuel and food production do not compete for resources, although this is not always the case.

A problem with the combustion of solid biomass fuels is that it emits considerable amounts of pollutants, such as particulates and polycyclic aromatic hydrocarbons. Even modern pellet boilers generate much more pollutants than oil or natural gas boilers. Pellets made from agricultural residues are usually worse than wood pellets, producing much larger emissions of dioxins and chlorophenols.

A derived fuel is biochar, which is produced by biomass pyrolysis. Biochar made from agricultural waste can substitute for wood charcoal. As wood stock becomes scarce, this alternative is gaining ground. In eastern Democratic Republic of Congo, for example, biomass briquettes are being marketed as an alternative to charcoal to protect Virunga National Park from deforestation associated with charcoal production.

Second-generation (Advanced) Biofuels

Second generation biofuels, also known as advanced biofuels, are fuels that can be manufactured from various types of biomass. Biomass is a wide-ranging term meaning any source of organic carbon that is renewed rapidly as part of the carbon cycle. Biomass is derived from plant materials, but can also include animal materials.

First generation biofuels are made from the sugars and vegetable oils found in arable crops, which can be easily extracted using conventional technology. In comparison, second generation biofuels are made from lignocellulosic biomass or woody crops, agricultural residues or waste. This makes it more difficult to extract the required fuel. A series of physical and chemical treatments might be required to convert lignocellulosic biomass to liquid fuels suitable for transportation.

Sustainable Biofuels

Biofuels in the form of liquid fuels derived from plant materials are entering the market, driven mainly by the perception that they reduce climate gas emissions, and also by factors such as oil price spikes and the need for increased energy security. However, many of the biofuels that are currently being supplied have been criticised for their adverse impacts on the natural environment, food security, and land use. In 2008, the Nobel-prize winning chemist Paul J. Crutzen published findings that the release of nitrous oxide (N_2O) emissions in the production of biofuels means that overall they contribute more to global warming than the fossil fuels they replace.

The challenge is to support biofuel development, including the development of new cellulosic technologies, with responsible policies and economic instruments to help ensure that biofuel commercialization is sustainable. Responsible commercialization of biofuels represents an opportunity to enhance sustainable economic prospects in Africa, Latin America and Asia.

According to the Rocky Mountain Institute, sound biofuel production practices would not hamper food and fibre production, nor cause water or environmental problems, and would enhance soil fertility. The selection of land on which to grow the feedstocks is a critical component of the ability of biofuels to deliver sustainable solutions. A key consideration is the minimisation of biofuel competition for prime cropland.

Biofuels by Region

Bio Diesel Powered Fast Attack Craft Of Indian Navy patrolling during IFR 2016.The green bands on the vessels are indicative of the fact that the vessels are powered by bio-diesel

There are international organizations such as IEA Bioenergy, established in 1978 by the OECD International Energy Agency (IEA), with the aim of improving cooperation and information exchange between countries that have national programs in bioenergy research, development and deployment. The UN International Biofuels Forum is formed by Brazil, China, India, Pakistan, South Africa, the United States and the European Commission. The world leaders in biofuel development and use are Brazil, the United States, France, Sweden and Germany. Russia also has 22% of world's forest, and is a big biomass (solid biofuels) supplier. In 2010, Russian pulp and paper maker, Vyborgskaya Cellulose, said they would be producing pellets that can be used in heat and electricity generation from its plant in Vyborg by the end of the year. The plant will eventually produce about 900,000 tons of pellets per year, making it the largest in the world once operational.

Biofuels currently make up 3.1% of the total road transport fuel in the UK or 1,440 million litres. By 2020, 10% of the energy used in UK road and rail transport must come from renewable sources – this is the equivalent of replacing 4.3 million tonnes of fossil oil each year. Conventional biofuels are likely to produce between 3.7 and 6.6% of the energy needed in road and rail transport, while advanced biofuels could meet up to 4.3% of the UK's renewable transport fuel target by 2020.

Air Pollution

Biofuels are different from fossil fuels in regard to greenhouse gases but are similar to fossil fuels in that biofuels contribute to air pollution. Burning produces airborne carbon particulates, carbon monoxide and nitrous oxides. The WHO estimates 3.7 million premature deaths worldwide in 2012 due to air pollution. Brazil burns significant amounts of ethanol biofuel. Gas chromatograph studies were performed of ambient air in São Paulo, Brazil, and compared to Osaka, Japan, which does not burn ethanol fuel. Atmospheric Formaldehyde was 160% higher in Brazil, and Acetaldehyde was 260% higher.

Debates Regarding the Production and use of Biofuel

There are various social, economic, environmental and technical issues with biofuel production and use, which have been discussed in the popular media and scientific journals. These include the effect of moderating oil prices, the "food vs fuel" debate, food prices, poverty reduction potential, energy ratio, energy requirements, carbon emissions levels, sustainable biofuel production, deforestation and soil erosion, loss of biodiversity, impact on water resources, the possible modifications necessary to run the engine on biofuel, as well as energy balance and efficiency. The International Resource Panel, which provides independent scientific assessments and expert advice on a variety of resource-related themes, assessed the issues relating to biofuel use in its first report *Towards sustainable production and use of resources: Assessing Biofuels*. "Assessing Biofuels" outlined the wider and interrelated factors that need to be considered when deciding on the relative merits of pursuing one biofuel over another. It concluded that not all biofuels perform equally in terms of their impact on climate, energy security and ecosystems, and suggested that environmental and social impacts need to be assessed throughout the entire life-cycle.

Another issue with biofuel use and production is the US has changed mandates many times because the production has been taking longer than expected. The Renewable Fuel Standard (RFS) set by congress for 2010 was pushed back to at best 2012 to produce 100 million gallons of pure ethanol (not blended with a fossil fuel).

Current Research

Research is ongoing into finding more suitable biofuel crops and improving the oil yields of these crops. Using the current yields, vast amounts of land and fresh water would be needed to produce enough oil to completely replace fossil fuel usage. It would require twice the land area of the US to be devoted to soybean production, or two-thirds to be devoted to rapeseed production, to meet current US heating and transportation needs.

Specially bred mustard varieties can produce reasonably high oil yields and are very useful in crop rotation with cereals, and have the added benefit that the meal left over after the oil has been pressed out can act as an effective and biodegradable pesticide.

The NFESC, with Santa Barbara-based Biodiesel Industries, is working to develop biofuels technologies for the US navy and military, one of the largest diesel fuel users in the world. A group of Spanish developers working for a company called Ecofasa announced a new biofuel made from trash. The fuel is created from general urban waste which is treated by bacteria to produce fatty acids, which can be used to make biofuels.

Ethanol Biofuels (Bioethanol)

As the primary source of biofuels in North America, many organizations are conducting research in the area of ethanol production. The National Corn-to-Ethanol Research Center (NCERC) is a research division of Southern Illinois University Edwardsville dedicated solely to ethanol-based biofuel research projects. On the federal level, the USDA conducts a large amount of research regarding ethanol production in the United States. Much of this research is targeted toward the effect of ethanol production on domestic food markets. A division of the U.S. Department of Energy, the National Renewable Energy Laboratory (NREL), has also conducted various ethanol research projects, mainly in the area of cellulosic ethanol.

Cellulosic ethanol commercialization is the process of building an industry out of methods of turning cellulose-containing organic matter into fuel. Companies, such as Iogen, POET, and Abengoa, are building refineries that can process biomass and turn it into bioethanol. Companies, such as Diversa, Novozymes, and Dyadic, are producing enzymes that could enable a cellulosic ethanol future. The shift from food crop feedstocks to waste residues and native grasses offers significant opportunities for a range of players, from farmers to biotechnology firms, and from project developers to investors.

As of 2013, the first commercial-scale plants to produce cellulosic biofuels have begun operating. Multiple pathways for the conversion of different biofuel feedstocks are being used. In the next few years, the cost data of these technologies operating at commercial scale, and their relative performance, will become available. Lessons learnt will lower the costs of the industrial processes involved.

In parts of Asia and Africa where drylands prevail, sweet sorghum is being investigated as a potential source of food, feed and fuel combined. The crop is particularly suitable for growing in arid conditions, as it only extracts one seventh of the water used by sugarcane. In India, and other places, sweet sorghum stalks are used to produce biofuel by squeezing the juice and then fermenting into ethanol.

A study by researchers at the International Crops Research Institute for the Semi-Arid Tropics (ICRISAT) found that growing sweet sorghum instead of grain sorghum could increase farmers incomes by US$40 per hectare per crop because it can provide fuel in addition to food and animal feed. With grain sorghum currently grown on over 11 million hectares (ha) in Asia and on 23.4 million ha in Africa, a switch to sweet sorghum could have a considerable economic impact.

Algae Biofuels

From 1978 to 1996, the US NREL experimented with using algae as a biofuels source in the "Aquatic Species Program". A self-published article by Michael Briggs, at the UNH Biofuels Group, offers estimates for the realistic replacement of all vehicular fuel with biofuels by using algae that have a natural oil content greater than 50%, which Briggs suggests can be grown on algae ponds at wastewater treatment plants. This oil-rich algae can then be extracted from the system and processed into biofuels, with the dried remainder further reprocessed to create ethanol. The production of algae to harvest oil for biofuels has not yet been undertaken on a commercial scale, but feasibility studies have been conducted to arrive at the above yield estimate. In addition to its projected high yield, algaculture – unlike crop-based biofuels – does not entail a decrease in food production, since it requires neither farmland nor fresh water. Many companies are pursuing algae bioreactors for various purposes, including scaling up biofuels production to commercial levels. Prof. Rodrigo E. Teixeira from the University of Alabama in Huntsville demonstrated the extraction of biofuels lipids from wet algae using a simple and economical reaction in ionic liquids.

Jatropha

Several groups in various sectors are conducting research on *Jatropha curcas*, a poisonous shrub-like tree that produces seeds considered by many to be a viable source of biofuels feedstock oil. Much of this research focuses on improving the overall per acre oil yield of Jatropha through advancements in genetics, soil science, and horticultural practices.

SG Biofuels, a San Diego-based jatropha developer, has used molecular breeding and biotechnology to produce elite hybrid seeds that show significant yield improvements over first-generation varieties. SG Biofuels also claims additional benefits have arisen from such strains, including improved flowering synchronicity, higher resistance to pests and diseases, and increased cold-weather tolerance.

Plant Research International, a department of the Wageningen University and Research Centre in the Netherlands, maintains an ongoing Jatropha Evaluation Project that examines the feasibility of large-scale jatropha cultivation through field and laboratory experiments. The Center for Sustainable Energy Farming (CfSEF) is a Los Angeles-based nonprofit research organization dedicated to jatropha research in the areas of plant science, agronomy, and horticulture. Successful exploration of these disciplines is projected to increase jatropha farm production yields by 200-300% in the next 10 years.

Fungi

A group at the Russian Academy of Sciences in Moscow, in a 2008 paper, stated they had isolated large amounts of lipids from single-celled fungi and turned it into biofuels in an economically efficient manner. More research on this fungal species, *Cunninghamella japonica*, and others, is likely to appear in the near future. The recent discovery of a variant of the fungus *Gliocladium roseum* (later renamed Ascocoryne sarcoides) points toward the production of so-called myco-diesel from cellulose. This organism was recently discovered in the rainforests of northern Patagonia, and has the unique capability of converting cellulose into medium-length hydrocarbons typically found in diesel fuel. Many other fungi that can degrade cellulose and other polymers have been observed to produce molecules that are currently being engineered using organisms from other kingdoms, suggesting that fungi may play a large role in the bio-production of fuels in the future (reviewed in).

Animal Gut Bacteria

Microbial gastrointestinal flora in a variety of animals have shown potential for the production of biofuels. Recent research has shown that TU-103, a strain of *Clostridium* bacteria found in Zebra feces, can convert nearly any form of cellulose into butanol fuel. Microbes in panda waste are being investigated for their use in creating biofuels from bamboo and other plant materials. There has also been substantial research into the technology of using the gut microbiomes of wood-feeding insects for the conversion of lignocellulotic material into biofuel.

Greenhouse Gas Emissions

Some scientists have expressed concerns about land-use change in response to greater demand for crops to use for biofuel and the subsequent carbon emissions. The payback period, that is, the time it will take biofuels to pay back the carbon debt they acquire due to land-use change, has been estimated to be between 100 and 1000 years, depending on the specific instance and location of land-use change. However, no-till practices combined with cover-crop practices can reduce the payback period to three years for grassland conversion and 14 years for forest conversion.

A study conducted in the Tocantis State, in northern Brazil, found that many families were cutting down forests in order to produce two conglomerates of oilseed plants, the J. curcas (JC group) and the R. communis (RC group). This region is composed of 15% Amazonian rainforest with high biodiversity, and 80% Cerrado forest with lower biodiversity. During the study, the farmers that planted the JC group released over 2193 Mg CO_2, while losing 53-105 Mg CO_2 sequestration from deforestation; and the RC group farmers released 562 Mg CO_2, while losing 48-90 Mg CO_2 to be sequestered from forest depletion. The production of these types of biofuels not only led into an increased emission of carbon dioxide, but also to lower efficiency of forests to absorb the gases that these farms were emitting. This has to do with the amount of fossil fuel the production of fuel crops involves. In addition, the intensive use of monocropping agriculture requires large amounts of water irrigation, as well as of fertilizers, herbicides and pesticides. This does not only lead to poisonous chemicals to disperse on water runoff, but also to the emission of nitrous oxide (NO_2) as a fertilizer byproduct, which is three hundred times more efficient in producing a greenhouse effect than carbon dioxide (CO_2).

Converting rainforests, peatlands, savannas, or grasslands to produce food crop–based biofuels in Brazil, Southeast Asia, and the United States creates a "biofuel carbon debt" by releasing 17 to 420 times more CO_2 than the annual greenhouse gas (GHG) reductions that these biofuels would provide by displacing fossil fuels. Biofuels made from waste biomass or from biomass grown on abandoned agricultural lands incur little to no carbon debt.

Water Use

In addition to crop growth requiring water, biofuel facilities require significant process water.

Biogas

Pipes carrying biogas (foreground), natural gas and condensate

Biogas typically refers to a mixture of different gases produced by the breakdown of organic matter in the absence of oxygen. Biogas can be produced from raw materials such as agricultural waste, manure, municipal waste, plant material, sewage, green waste or food waste. Biogas is a renewable energy source.

Biogas can be produced by anaerobic digestion with anaerobic organisms, which digest material inside a closed system, or fermentation of biodegradable materials.

Biogas is primarily methane (CH_4) and carbon dioxide (CO_2) and may have small amounts of hydrogen sulfide (H_2S), moisture and siloxanes. The gases methane, hydrogen, and carbon monoxide (CO) can be combusted or oxidized with oxygen. This energy release allows biogas to be used as a fuel; it can be used for any heating purpose, such as cooking. It can also be used in a gas engine to convert the energy in the gas into electricity and heat.

Biogas can be compressed, the same way natural gas is compressed to CNG, and used to power motor vehicles. In the UK, for example, biogas is estimated to have the potential to replace around 17% of vehicle fuel. It qualifies for renewable energy subsidies in some parts of the world. Biogas can be cleaned and upgraded to natural gas standards, when it becomes bio-methane. Biogas is considered to be a renewable resource because its production-and-use cycle is continuous, and it generates no net carbon dioxide. Organic material grows, is converted and used and then regrows in a continually repeating cycle. From a carbon perspective, as much carbon dioxide is absorbed from the atmosphere in the growth of the primary bio-resource as is released when the material is ultimately converted to energy.

Production

Biogas production in rural Germany

Biogas is produced as landfill gas (LFG), which is produced by the breakdown of Biodegradable waste inside a landfill due to chemical reactions and microbes, or as digested gas, produced inside an anaerobic digester. A *biogas plant* is the name often given to an anaerobic digester that treats farm wastes or energy crops. It can be produced using anaerobic digesters (air-tight tanks with different configurations). These plants can be fed with energy crops such as maize silage or biodegradable wastes including sewage sludge and food waste. During the process, the microorganisms transform biomass waste into biogas (mainly methane and carbon dioxide) and digestate. The biogas is a renewable energy that can be used for heating, electricity, and many other operations that use a reciprocating internal combustion engine, such as GE Jenbacher or Caterpillar gas engines. Other internal combustion engines such as gas turbines are suitable for the conversion of biogas into both electricity and heat. The digestate is the remaining inorganic matter that was not transformed into biogas. It can be used as an agricultural fertiliser.

There are two key processes: mesophilic and thermophilic digestion which is dependent on temperature. In experimental work at University of Alaska Fairbanks, a 1000-litre digester using psychrophiles harvested from "mud from a frozen lake in Alaska" has produced 200–300 liters of methane per day, about 20%–30% of the output from digesters in warmer climates.

Dangers

The dangers of biogas are mostly similar to those of natural gas, but with an additional risk from the toxicity of its hydrogen sulfide fraction. Biogas can be explosive when mixed in the ratio of one part biogas to 8-20 parts air. Special safety precautions have to be taken for entering an empty biogas digester for maintenance work.

It is important that a biogas system never has negative pressure as this could cause an explosion. Negative gas pressure can occur if too much gas is removed or leaked; Because of this biogas should not be used at pressures below one column inch of water, measured by a pressure gauge.

Frequent smell checks must be performed on a biogas system. If biogas is smelled anywhere windows and doors should be opened immediately. If there is a fire the gas should be shut off at the gate valve of the biogas system.

Landfill Gas

Landfill gas is produced by wet organic waste decomposing under anaerobic conditions in a biogas.

The waste is covered and mechanically compressed by the weight of the material that is deposited above. This material prevents oxygen exposure thus allowing anaerobic microbes to thrive. Biogas builds up and is slowly released into the atmosphere if the site has not been engineered to capture the gas. Landfill gas released in an uncontrolled way can be hazardous since it can become explosive when it escapes from the landfill and mixes with oxygen. The lower explosive limit is 5% methane and the upper is 15% methane.

The methane in biogas is 20 times more potent a greenhouse gas than carbon dioxide. Therefore, uncontained landfill gas, which escapes into the atmosphere may significantly contribute to the effects of global warming. In addition, volatile organic compounds (VOCs) in landfill gas contribute to the formation of photochemical smog.

Technical

Biochemical oxygen demand (BOD) is a measure of the amount of oxygen required by aerobic micro-organisms to decompose the organic matter in a sample of water. Knowing the energy density of the material being used in the biodigester as well as the BOD for the liquid discharge allows for the calculation of the daily energy output from a biodigester.

Another term related to biodigesters is effluent dirtiness, which tells how much organic material there is per unit of biogas source. Typical units for this measure are in mg BOD/litre. As an example, effluent dirtiness can range between 800–1200 mg BOD/litre in Panama.

From 1 kg of decommissioned kitchen bio-waste, 0.45 m^3 of biogas can be obtained. The price for collecting biological waste from households is approximately €70 per ton.

Composition

Typical composition of biogas		
Compound	**Formula**	**%**
Methane	CH_4	50–75
Carbon dioxide	CO_2	25–50
Nitrogen	N_2	0–10
Hydrogen	H_2	0–1
Hydrogen sulfide	H_2S	0–3
Oxygen	O_2	0–0.5

The composition of biogas varies depending upon the origin of the anaerobic digestion process. Landfill gas typically has methane concentrations around 50%. Advanced waste treatment technologies can produce biogas with 55%–75% methane, which for reactors with free liquids can be increased to 80%-90% methane using in-situ gas purification techniques. As produced, biogas contains water vapor. The fractional volume of water vapor is a function of biogas temperature; correction of measured gas volume for water vapor content and thermal expansion is easily done via simple mathematics which yields the standardized volume of dry biogas.

In some cases, biogas contains siloxanes. They are formed from the anaerobic decomposition of materials commonly found in soaps and detergents. During combustion of biogas containing siloxanes, silicon is released and can combine with free oxygen or other elements in the combustion gas. Deposits are formed containing mostly silica (SiO_2) or silicates (Si_xO_y) and can contain calcium, sulfur, zinc, phosphorus. Such *white mineral* deposits accumulate to a surface thickness of several millimeters and must be removed by chemical or mechanical means.

Practical and cost-effective technologies to remove siloxanes and other biogas contaminants are available.

For 1000 kg (wet weight) of input to a typical biodigester, total solids may be 30% of the wet weight while volatile suspended solids may be 90% of the total solids. Protein would be 20% of the volatile solids, carbohydrates would be 70% of the volatile solids, and finally fats would be 10% of the volatile solids.

Benefits of Manure Derived Biogas

High levels of methane are produced when manure is stored under anaerobic conditions. During storage and when manure has been applied to the land, nitrous oxide is also produced as a byproduct of the denitrification process. Nitrous oxide (N_2O) is 320 times more aggressive as a greenhouse gas than carbon dioxide and methane 25 times more than carbon dioxide.

By converting cow manure into methane biogas via anaerobic digestion, the millions of cattle in the United States would be able to produce 100 billion kilowatt hours of electricity, enough to power millions of homes across the United States. In fact, one cow can produce enough manure in one day to generate 3 kilowatt hours of electricity; only 2.4 kilowatt hours of electricity are needed to power a single 100-watt light bulb for one day. Furthermore, by converting cattle manure into methane biogas instead of letting it decompose, global warming gases could be reduced by 99 million metric tons or 4%.

Applications

A biogas bus in Linköping, Sweden

Biogas can be used for electricity production on sewage works, in a CHP gas engine, where the waste heat from the engine is conveniently used for heating the digester; cooking; space heating; water heating; and process heating. If compressed, it can replace compressed natural gas for use in vehicles, where it can fuel an internal combustion engine or fuel cells and is a much more effective displacer of carbon dioxide than the normal use in on-site CHP plants.

Biogas Upgrading

Raw biogas produced from digestion is roughly 60% methane and 29% CO_2 with trace elements of H_2S; it is not of high enough quality to be used as fuel gas for machinery. The corrosive nature of H_2S alone is enough to destroy the internals of a plant.

Methane in biogas can be concentrated via a biogas upgrader to the same standards as fossil natural gas, which itself has to go through a cleaning process, and becomes *biomethane*. If the local gas network allows, the producer of the biogas may use their distribution networks. Gas must be very clean to reach pipeline quality and must be of the correct composition for the distribution network to accept. Carbon dioxide, water, hydrogen sulfide, and particulates must be removed if present.

There are four main methods of upgrading: water washing, pressure swing adsorption, selexol adsorption, and amine gas treating. In addition to these, the use of membrane separation technology for biogas upgrading is increasing, and there are already several plants operating in Europe and USA.

The most prevalent method is water washing where high pressure gas flows into a column where the carbon dioxide and other trace elements are scrubbed by cascading water running counter-flow to the gas. This arrangement could deliver 98% methane with manufacturers guaranteeing maximum 2% methane loss in the system. It takes roughly between 3% and 6% of the total energy output in gas to run a biogas upgrading system.

Biogas Gas-grid Injection

Gas-grid injection is the injection of biogas into the methane grid (natural gas grid). Injections includes biogas until the breakthrough of micro combined heat and power two-thirds of all the energy produced by biogas power plants was lost (the heat), using the grid to transport the gas to customers, the electricity and the heat can be used for on-site generation resulting in a reduction of losses in the transportation of energy. Typical energy losses in natural gas transmission systems range from 1% to 2%. The current energy losses on a large electrical system range from 5% to 8%.

Biogas in Transport

"Biogaståget Amanda" ("Amanda the Biogas Train") train near Linköping station, Sweden

If concentrated and compressed, it can be used in vehicle transportation. Compressed biogas is becoming widely used in Sweden, Switzerland, and Germany. A biogas-powered train, named Biogaståget Amanda (The Biogas Train Amanda), has been in service in Sweden since 2005. Biogas powers automobiles. In 1974, a British documentary film titled *Sweet as a Nut* detailed the biogas production process from pig manure and showed how it fueled a custom-adapted combustion engine. In 2007, an estimated 12,000 vehicles were being fueled with upgraded biogas worldwide, mostly in Europe.

Measuring in Biogas Environments

Biogas is part of the wet gas and condensing gas (or air) category that includes mist or fog in the gas stream. The mist or fog is predominately water vapor that condenses on the sides of pipes or stacks throughout the gas flow. Biogas environments include wastewater digesters, landfills, and animal feeding operations (covered livestock lagoons).

Ultrasonic flow meters are one of the few devices capable of measuring in a biogas atmosphere. Most of thermal flow meters are unable to provide reliable data because the moisture causes steady high flow readings and continuous flow spiking, although there are single-point insertion thermal mass flow meters capable of accurately monitoring biogas flows with minimal pressure drop. They can handle moisture variations that occur in the flow stream because of daily and seasonal temperature fluctuations, and account for the moisture in the flow stream to produce a dry gas value.

Legislation

European Union

The European Union has legislation regarding waste management and landfill sites called the Landfill Directive.

Countries such as the United Kingdom and Germany now have legislation in force that provides farmers with long-term revenue and energy security.

United States

The United States legislates against landfill gas as it contains VOCs. The United States Clean Air Act and Title 40 of the Code of Federal Regulations (CFR) requires landfill owners to estimate the quantity of non-methane organic compounds (NMOCs) emitted. If the estimated NMOC emissions exceeds 50 tonnes per year, the landfill owner is required to collect the gas and treat it to remove the entrained NMOCs. Treatment of the landfill gas is usually by combustion. Because of the remoteness of landfill sites, it is sometimes not economically feasible to produce electricity from the gas.

Global Developments

United States

With the many benefits of biogas, it is starting to become a popular source of energy and is starting to be used in the United States more. In 2003, the United States consumed 147 trillion BTU of energy from "landfill gas", about 0.6% of the total U.S. natural gas consumption. Methane biogas derived from cow manure is being tested in the U.S. According to a 2008 study, collected by the *Science and Children* magazine, methane biogas from cow manure would be sufficient to produce 100 billion kilowatt hours enough to power millions of homes across America. Furthermore, methane biogas has been tested to prove that it can reduce 99 million metric tons of greenhouse gas emissions or about 4% of the greenhouse gases produced by the United States.

In Vermont, for example, biogas generated on dairy farms was included in the CVPS Cow Power program. The program was originally offered by Central Vermont Public Service Corporation as a voluntary tariff and now with a recent merger with Green Mountain Power is now the GMP Cow Power Program. Customers can elect to pay a premium on their electric bill, and that premium is passed directly to the farms in the program. In Sheldon, Vermont, Green Mountain Dairy has provided renewable energy as part of the Cow Power program. It started when the brothers who own the farm, Bill and Brian Rowell, wanted to address some of the manure management challenges faced by dairy farms, including manure odor, and nutrient availability for the crops they need to grow to feed the animals. They installed an anaerobic digester to process the cow and milking center waste from their 950 cows to produce renewable energy, a bedding to replace sawdust, and a plant-friendly fertilizer. The energy and environmental attributes are sold to the GMP Cow Power program. On average, the system run by the Rowells produces enough electricity to power 300 to 350 other homes. The generator capacity is about 300 kilowatts.

In Hereford, Texas, cow manure is being used to power an ethanol power plant. By switching to methane biogas, the ethanol power plant has saved 1000 barrels of oil a day. Over all, the power plant has reduced transportation costs and will be opening many more jobs for future power plants that will rely on biogas.

In Oakley, Kansas, an ethanol plant considered to be one of the largest biogas facilities in North America is using Integrated Manure Utilization System "IMUS" to produce heat for its boilers by utilizing feedlot manure, municipal organics and ethanol plant waste. At full capacity the plant is expected to replace 90% of the fossil fuel used in the manufacturing process of ethanol.

Europe

The level of development varies greatly in Europe. While countries such as Germany, Austria and Sweden are fairly advanced in their use of biogas, there is a vast potential for this renewable energy source in the rest of the continent, especially in Eastern Europe. Different legal frameworks, education schemes and the availability of technology are among the prime reasons behind this untapped potential. Another challenge for the further progression of biogas has been negative public perception.

In February 2009, the European Biogas Association (EBA) was founded in Brussels as a non-profit organisation to promote the deployment of sustainable biogas production and use in Europe. EBA's strategy defines three priorities: establish biogas as an important part of Europe's energy mix, promote source separation of household waste to increase the gas potential, and support the production of biomethane as vehicle fuel. In July 2013, it had 60 members from 24 countries across Europe.

UK

As of September 2013, there are about 130 non-sewage biogas plants in the UK. Most are on-farm, and some larger facilities exist off-farm, which are taking food and consumer wastes.

On 5 October 2010, biogas was injected into the UK gas grid for the first time. Sewage from over 30,000 Oxfordshire homes is sent to Didcot sewage treatment works, where it is treated in an anaerobic digestor to produce biogas, which is then cleaned to provide gas for approximately 200 homes.

In 2015 the Green-Energy company Ecotricity announced their plans to build three grid-injecting digesters.

Germany

Germany is Europe's biggest biogas producer and the market leader in biogas technology. In 2010 there were 5,905 biogas plants operating throughout the country: Lower Saxony, Bavaria, and the eastern federal states are the main regions. Most of these plants are employed as power plants. Usually the biogas plants are directly connected with a CHP which produces electric power by burning the bio methane. The electrical power is then fed into the public power grid. In 2010, the total installed electrical capacity of these power plants was 2,291 MW. The electricity supply was approximately 12.8 TWh, which is 12.6% of the total generated renewable electricity.

Biogas in Germany is primarily extracted by the co-fermentation of energy crops (called 'NawaRo', an abbreviation of *nachwachsende Rohstoffe*, German for renewable resources) mixed with manure. The main crop used is corn. Organic waste and industrial and agricultural residues such as waste from the food industry are also used for biogas generation. In this respect, biogas production in Germany differs significantly from the UK, where biogas generated from landfill sites is most common.

Biogas production in Germany has developed rapidly over the last 20 years. The main reason is the legally created frameworks. Government support of renewable energy started in 1991 with the Electricity Feed-in Act (*StrEG*). This law guaranteed the producers of energy from renewable sources the feed into the public power grid, thus the power companies were forced to take all produced energy from independent private producers of green energy. In 2000 the Electricity Feed-in Act was replaced by the Renewable Energy Sources Act (*EEG*). This law even guaranteed a fixed compensation for the produced electric power over 20 years. The amount of around 8 ¢/kWh gave farmers the opportunity to become energy suppliers and gain a further source of income.

The German agricultural biogas production was given a further push in 2004 by implementing the so-called NawaRo-Bonus. This is a special payment given for the use of renewable resources, that is, energy crops. In 2007 the German government stressed its intention to invest further effort and support in improving the renewable energy supply to provide an answer on growing climate challenges and increasing oil prices by the 'Integrated Climate and Energy Programme'.

This continual trend of renewable energy promotion induces a number of challenges facing the management and organisation of renewable energy supply that has also several impacts on the biogas production. The first challenge to be noticed is the high area-consuming of the biogas electric power supply. In 2011 energy crops for biogas production consumed an area of circa 800,000 ha in Germany. This high demand of agricultural areas generates new competitions with the food industries that did not exist hitherto. Moreover, new industries and markets were created in predominately rural regions entailing different new players with an economic, political and civil background. Their influence and acting has to be governed to gain all advantages this new source of energy is offering. Finally biogas will furthermore play an important role in the German renewable energy supply if good governance is focused.

Indian Subcontinent

Biogas in India has been traditionally based on dairy manure as feed stock and these "gobar" gas plants have been in operation for a long period of time, especially in rural India. In the last 2-3 decades, research organisations with a focus on rural energy security have enhanced the design of the systems resulting in newer efficient low cost designs such as the Deenabandhu model.

The Deenabandhu Model is a new biogas-production model popular in India. (*Deenabandhu* means "friend of the helpless.") The unit usually has a capacity of 2 to 3 cubic metres. It is constructed using bricks or by a ferrocement mixture. In India, the brick model costs slightly more than the ferrocement model; however, India's Ministry of New and Renewable Energy offers some subsidy per model constructed.

LPG (Liquefied Petroleum Gas) is a key source of cooking fuel in urban India and its prices have been increasing along with the global fuel prices. Also the heavy subsidies provided by the successive governments in promoting LPG as a domestic cooking fuel has become a financial burden renewing the focus on biogas as a cooking fuel alternative in urban establishments. This has led to the development of prefabricated digester for modular deployments as compared to RCC and cement structures which take a longer duration to construct. Renewed focus on process technology like the Biourja process model has enhanced the stature of medium and large scale anaerobic digester in India as a potential alternative to LPG as primary cooking fuel.

In India, Nepal, Pakistan and Bangladesh biogas produced from the anaerobic digestion of manure in small-scale digestion facilities is called gobar gas; it is estimated that such facilities exist in over 2 million households in India, 50,000 in Bangladesh and thousands in Pakistan, particularly North Punjab, due to the thriving population of livestock. The digester is an airtight circular pit made of concrete with a pipe connection. The manure is directed to the pit, usually straight from the cattle shed. The pit is filled with a required quantity of wastewater. The gas pipe is connected to the kitchen fireplace through control valves. The combustion of this biogas has very little odour or smoke. Owing to simplicity in implementation and use of cheap raw materials in villages, it is one of the most environmentally sound energy sources for rural needs. One type of these system is the Sintex Digester. Some designs use vermiculture to further enhance the slurry produced by the biogas plant for use as compost.

To create awareness and associate the people interested in biogas, the Indian Biogas Association was formed. It aspires to be a unique blend of nationwide operators, manufacturers and planners of biogas plants, and representatives from science and research. The association was founded in 2010 and is now ready to start mushrooming. Its motto is "propagating Biogas in a sustainable way".

In Pakistan, the Rural Support Programmes Network is running the Pakistan Domestic Biogas Programme which has installed 5,360 biogas plants and has trained in excess of 200 masons on the technology and aims to develop the Biogas Sector in Pakistan.

In Nepal, the government provides subsidies to build biogas plant at home.

China

The Chinese have experimented with the applications of biogas since 1958. Around 1970, China had installed 6,000,000 digesters in an effort to make agriculture more efficient. During the last years the technology has met high growth rates. This seems to be the earliest developments in generating biogas from agricultural waste.

In Developing Nations

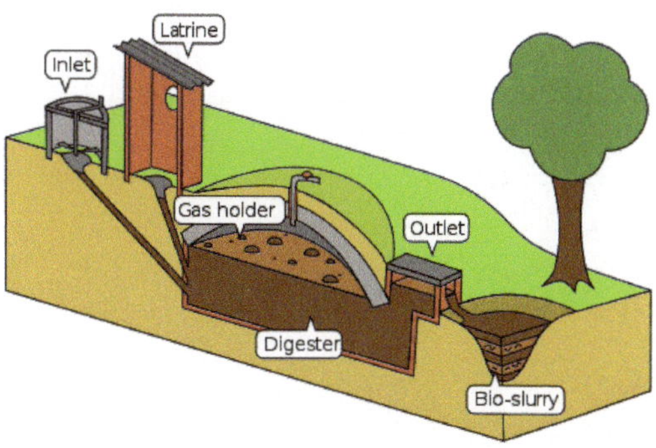

Simple sketch of household biogas plant

Domestic biogas plants convert livestock manure and night soil into biogas and slurry, the fermented manure. This technology is feasible for small-holders with livestock producing 50 kg manure per day, an equivalent of about 6 pigs or 3 cows. This manure has to be collectable to mix it with water and feed it into the plant. Toilets can be connected. Another precondition is the temperature that affects the fermentation process. With an optimum at 36 C° the technology especially applies for those living in a (sub) tropical climate. This makes the technology for small holders in developing countries often suitable.

Depending on size and location, a typical brick made fixed dome biogas plant can be installed at the yard of a rural household with the investment between US$300 to $500 in Asian countries and up to $1400 in the African context. A high quality biogas plant needs minimum maintenance costs and can produce gas for at least 15–20 years without major problems and re-investments. For the user, biogas provides clean cooking energy, reduces indoor air pollution, and reduces the time needed for traditional biomass collection, especially for women and children. The slurry is a clean organic fertilizer that potentially increases agricultural productivity.

Domestic biogas technology is a proven and established technology in many parts of the world, especially Asia. Several countries in this region have embarked on large-scale programmes on domestic biogas, such as China and India.

The Netherlands Development Organisation, SNV, supports national programmes on domestic biogas that aim to establish commercial-viable domestic biogas sectors in which local companies market, install and service biogas plants for households. In Asia, SNV is working in Nepal, Vietnam, Bangladesh, Bhutan, Cambodia, Lao PDR, Pakistan and Indonesia, and in Africa; Rwanda, Senegal, Burkina Faso, Ethiopia, Tanzania, Uganda, Kenya, Benin and Cameroon.

In South Africa a prebuilt Biogas system is manufactured and sold. One key feature is that installation requires less skill and is quicker to install as the digester tank is premade plastic.

Society and Culture

In the 1985 Australian film *Mad Max Beyond Thunderdome* the post-apocalyptic settlement Barter town is powered by a central biogas system based upon a piggery. As well as providing electricity, methane is used to power Barter's vehicles.

"Cow Town", written in the early 1940s, discuss the travails of a city vastly built on cow manure and the hardships brought upon by the resulting methane biogas. Carter McCormick, an engineer from a town outside the city, is sent in to figure out a way to utilize this gas to help power, rather than suffocate, the city.

References

- "Valcent Products Inc. Develops "Clean Green" Vertical Bio-Reactor". Valcent Products. Archived from the original on 18 June 2008. Retrieved 2008-07-09
- R. E. Teixeira (2012). "Energy-efficient extraction of fuel and chemical feedstocks from algae". Green Chemistry. 14 (2): 419–427. doi:10.1039/C2GC16225C
- Koyama, Tadashiro (1963). "Gaseous metabolism in lake sediments and paddy soils and the production of atmospheric methane and hydrogen". Journal of Geophysical Research. 68 (13): 3971–3973. doi:10.1029/JZ068i013p03971
- Thomas F. McGowan, Michael L. Brown, William S. Bulpitt, James L. Walsh, Jr. (Editors) 2009. Biomass and Alternate Fuel Systems: An Engineering and Economic Guide. ISBN 978-0-470-41028-8 Wiley, 280 pages

- "Press Conference Launching International Biofuels Forum". United Nations Department of Public Information. 2007-03-02. Retrieved 2008-01-15
- Hammerschlag, R (2006). "Ethanol's Energy Return on Investment: A Survey of the Literature 1999-Present". Environ. Sci. Technol. 40: 1744–50. doi:10.1021/es052024h
- Hill, D. T.; Barth, C. L. (1977). "A Dynamic Model for Simulation of Animal Waste Digestion". Journal (Water Pollution Control Federation). 49 (10): 2129–2143
- The National Academies Press (2008). "Water Issues of Biofuel Production Plants". The National Academies Press. Retrieved 18 June 2015
- B.N. Divakara; H.D. Upadhyaya; S.P. Wani; C.L. Laxmipathi Gowda (2010). "Biology and genetic improvement of Jatropha curcas L.: A review". Applied Energy. 87 (3): 732–742. doi:10.1016/j.apenergy.2009.07.013
- Zehnder, Alexander J. B. (1978). "Ecology of methane formation". In Mitchell, Ralph. Water pollution microbiology 2. New York: Wiley. pp. 349–376. ISBN 978-0-471-01902-2
- Briens, Cedric; Piskorz, Jan; Berruti, Franco (2008). "Biomass Valorization for Fuel and Chemicals Production -- A Review". International Journal of Chemical Reactor Engineering. 6: R2. doi:10.2202/1542-6580.1674
- Shah, Dhruti (5 October 2010). "Oxfordshire town sees human waste used to heat homes". BBC NEWS. Archived from the original on 5 October 2010. Retrieved 5 October 2010
- R. E. Teixeira (2012). "Energy-efficient extraction of fuel and chemical feedstocks from algae". Green Chemistry. 14 (2): 419–427. doi:10.1039/C2GC16225C

Permissions

All chapters in this book are published with permission under the Creative Commons Attribution Share Alike License or equivalent. Every chapter published in this book has been scrutinized by our experts. Their significance has been extensively debated. The topics covered herein carry significant information for a comprehensive understanding. They may even be implemented as practical applications or may be referred to as a beginning point for further studies.

We would like to thank the editorial team for lending their expertise to make the book truly unique. They have played a crucial role in the development of this book. Without their invaluable contributions this book wouldn't have been possible. They have made vital efforts to compile up to date information on the varied aspects of this subject to make this book a valuable addition to the collection of many professionals and students.

This book was conceptualized with the vision of imparting up-to-date and integrated information in this field. To ensure the same, a matchless editorial board was set up. Every individual on the board went through rigorous rounds of assessment to prove their worth. After which they invested a large part of their time researching and compiling the most relevant data for our readers.

The editorial board has been involved in producing this book since its inception. They have spent rigorous hours researching and exploring the diverse topics which have resulted in the successful publishing of this book. They have passed on their knowledge of decades through this book. To expedite this challenging task, the publisher supported the team at every step. A small team of assistant editors was also appointed to further simplify the editing procedure and attain best results for the readers.

Apart from the editorial board, the designing team has also invested a significant amount of their time in understanding the subject and creating the most relevant covers. They scrutinized every image to scout for the most suitable representation of the subject and create an appropriate cover for the book.

The publishing team has been an ardent support to the editorial, designing and production team. Their endless efforts to recruit the best for this project, has resulted in the accomplishment of this book. They are a veteran in the field of academics and their pool of knowledge is as vast as their experience in printing. Their expertise and guidance has proved useful at every step. Their uncompromising quality standards have made this book an exceptional effort. Their encouragement from time to time has been an inspiration for everyone.

The publisher and the editorial board hope that this book will prove to be a valuable piece of knowledge for students, practitioners and scholars across the globe.

Index